A Future for the Land

Organic Practice from a Global Perspective

Edited by

Philip Conford

A RESURGENCE BOOK

A Resurgence Book
First published in 1992 by
Green Books
Ford House, Hartland
Bideford, Devon
EX39 6EE

Composed in Garamond 10 on 12½
by Chris Fayers
Soldon, Devon, EX22 7PF

Printed by Biddles Ltd
Guildford, Surrey

British Library Cataloguing in Publication Data
A Future for the Land: Organic
Practice from a Global Perspective
I. Conford, Philip
338.1

ISBN 1-870098-49-8

CONTENTS

To Helen and Mark

ACKNOWLEDGEMENTS

Chiefly, of course, my thanks must go to the contributors to this book for their stimulating and well-informed chapters. All of them are busy people, and I much appreciate their generosity in setting aside time for the project. In particular I would like to thank Robert Waller, who offered valuable support and enjoyable hospitality during the period of the book's preparation. Julian Rose's help and suggestions during the planning of the book were similarly encouraging, and his hospitality was a convincing argument in favour of organic farming. Hugh Raven, former Research Assistant to Ron Davies, must also be mentioned, since he ensured that time was found in an MP's schedule for writing the chapter on Parliament's possible role in creating an organic future. Chapter 10 appears with the kind permission of John P. Reganold, Robert I. Papendick and James F. Parr, and of *Scientific American*, where it first appeared. I would also like to thank Clive Ponting for agreeing to provide a Foreword to the book which so clearly and incisively sets the chapters in their context.

I have been helped once again by the library staff at Chichester College of Technology; Miss Clare Likeman and Miss Christina Thomas have been most tolerant of the demands made on their time by my many requests for inter-library loans.

Finally I would like to thank John Elford, of Green Books, for all his assistance, and Satish Kumar, for conceiving the project and entrusting it to me for completion.

Foreword

For the last ten thousand years human societies have relied on a peculiar, artificial ecosystem for their survival—agriculture. In many respects it has been highly successful. It has enabled far more people to live than would a hunting and gathering system: the world's population has risen from about four million, when agriculture was adopted, to over five billion now. It has also led to everything we grandly call 'civilization'. Because a farming family produce far more food than they can consume, a surplus is available to feed all the industrial workers, bureaucrats, priests, artists, scientists, musicians, office workers and others that a modern society needs.

Agriculture, however, poses a major ecological problem. It destroys the natural climax ecosystem with all its diversity and balanced recycling processes, and substitutes a simplified system restricted to just a few crops and with the natural recycling mechanisms removed. Agriculture therefore depends on inputs from outside the system—manures and fertilizers—in order to keep up productivity. What most of the advocates of modern hi-tech farming forget is that until the last two centuries all agriculture depended on natural, organic inputs—animal manures and plant wastes. This system was highly successful, producing steady, if unspectacular, gains in output over the centuries as new techniques—like using legumes as fodder crops—were introduced and old ones subtly modified.

In the last ten thousand years agriculture has spread across the globe. About 11 per cent of the world's surface is now used for growing crops and about 25 per cent for grazing animals. There is now very little new land to bring into production. The result has been a vast increase in intensification in the last two centuries. This began in the nineteenth century with the large-scale use by Europe and North America of imported fertilizers—guano and phosphates. At first these extra inputs were natural products. It was only with the chemical revolution which began at the end of the nineteenth century that artificial products—at first fertilizers, but subsequently highly toxic compounds derived from some of the original military research into chemical warfare—were introduced as pesticides, herbicides and fungicides.

Modern hi-tech agriculture, particularly as practised in the industrialised world, is therefore completely dependent on a whole variety of artificial inputs. It is also dependent on a high degree of industrialisation—machinery, heated buildings and industrial processes. This is a fundamental departure from all previous practice. We are constantly told, by both governments and farmers' organisations, that this new form of agriculture is highly efficient. It is actually a highly inefficient way of producing food.

The most energy-efficient form of farming in the world is the rice paddy-field. It produces about fifty times as much energy as it consumes. Modern cereal farming produces only about twice as much energy as it consumes in the form of fertilisers, pesticides and machinery. Meat production is even worse—it consumes about two to three times the energy it produces. Modern agriculture is, moreover, becoming steadily less energy-efficient. In the 25 years after the Second World War the energy inputs into US agriculture increased by 400 per cent but yields only rose about 140 per cent. Overall the energy efficiency of US corn production has fallen by a half since 1915.

Modern industrialised agriculture is therefore travelling up a cul-de-sac. Increasing use of inputs that are themselves steadily rising in cost in order to get ever smaller rises in output cannot be continued indefinitely. Modern agriculture also has two other major side effects. Firstly, it is de-humanising—the number of people working on the land has been falling rapidly in the industrialised world during the last century, farms have been getting bigger, and the work involved has less and less to do with traditional agriculture. Animals are kept in overcrowded buildings, and the workers spend their days in tractor cabs surrounded by the noise of radio cassette players. Secondly, it is ruining the countryside for the rest of the community through the destruction of natural features such as woodland and hedgerows.

This drive to industrialise agriculture and increase output has also led to major ecological disasters because of the artifical nature of all agricultural systems. We see the results in salinisation, desertification and soil erosion. None of this is new though—many of the great civilizations of the past collapsed because they destroyed the basis of their agricultural systems. In southern Mesopotamia, the first literate societies emerged about 3000BC, but collapsed when continuous irrigation of poor soils led to salinisation and drastically reduced crop yields. The great granaries of the Roman empire in North Africa collapsed through desertification. The whole of the Mediterranean landscape is now a monument to overgrazing and soil erosion over many millenia.

These same problems are now facing us across the world. Globally we are losing more land to salinisation than irrigation is bringing into production. Cultivation of unsuitable land, such as the American Great Plains and Khazakstan, and bad practices elsewhere have produced massive soil erosion—a third of the top-soil of the United States has already been lost. Desertification is proceeding rapidly as well. The southern boundary of the Sahara desert moved south by over 120 miles in just seventeen years before 1975 and it is still advancing.

All of these disasters result from a human attitude to nature that involves ideas of mastery and domination—bending nature to our will. Time after time humans have come up against natural resistance to their grandiose schemes and been surprised at the consequences of pressing on regardless. Occasionally, though, humans have been able to find a way of life that has not resulted in environmental catastrophe. One such example is ancient Egypt where, although the political system was highly authoritarian, the agricultural system that supported the society was well adapted to the natural rhythm of the annual Nile flood. Instead of building large, artificial irrigation channels as in Mesopotamia, the ancient Egyptians adapted natural channels and reservoirs to produce a system that, despite many vicissitudes such as low and disastrously high floods, lasted for several millenia. It was the industrialised world that wrecked the system. First, the British introduced artifical irrigation which rapidly produced salinisation. Second, the building of successively higher dams at Aswan ruined the flood by removing most of the silt that produced the rich, fertile, constantly renewed soil of lower Egypt.

A more harmonious relationship between humans and agriculture is both possible and necessary. Farmers need to accept the limitations of the world in which they operate and the need to work with rather than against the system. It is often argued by the farmers, the agrochemical lobby, international organisations such as the FAO, and governments, that such an approach is hopelessly idealistic and optimistic and that the only way to deal with the problem of feeding the world's population is to go on industrialising farming. Such a view is based on a major fallacy, but one that is very welcome to vested interests around the globe.

There is no shortage of food in the world. But there is an appalling maldistribution of food. In a world where a cat in the United States eats more meat than the average citizen of the countries of Central America that produced the meat it is difficult to argue there is a world food shortage. The world's production of food and its distribution is controlled by the industrialised countries in their interest. Nearly twice as much food flows from the Third World to the industrialised world than in the other direction. It is not surprising therefore that the inhabitants of the industrialised countries eat a wide and varied diet, consume too many calories and artificial products of the food processing industry and suffer from the resulting diseases, while about half the world's population—two and a half billion people—live on an inadequate diet, 730 million people do not eat enough to constitute a minimum diet to maintain health and 40 million a year (half of them children) die of starvation.

The future of agriculture, like so much else in the world, depends on the attitudes people adopt. Alternatives, more in keeping with natural constraints, are available. But there is a limit to what individuals can do on their own. The policies of governments, multinational corporations and international organisations are even more important because they have far more power. If they continue in their current direction then the range and intensity of our problems will only get worse. We cannot argue that we did not know better or that we did not have the examples of the past to warn us.

Against this stark background, this new book *A Future for the Land* can make a significant contribution to bridging the gap between policy-formers and activists. With chapters by practitioners, politicians, writers on ecological issues, and academics, it brings together a wealth of material and a variety of perspectives on the criticam issues of land use and development that we face today.

Clive Ponting
Llandeilo, Dyfed
April 1992

Introduction

AN ORGANIC FUTURE. The implications of the phrase go beyond the image it may conjure up, of supermarket shelves crammed with fruit and vegetables approved by the Soil Association, allotment holders diligently applying methods seen on *All Muck and Magic?*, and East Anglian prairies reverting to small-scale mixed farms.

Certainly, a horticulture with a much-reduced reliance on chemicals, and an agriculture based on a sustainable approach to soil fertility, are necessary conditions of an organic future; but the organic as a concept has a long history that is independent of these immediate practical concerns. To know what general features an organic future would have, we need to examine the idea of the organic as a philosophical category, with scientific, aesthetic, and political implications.

The mainsprings of organic farming are the determination to see things as a whole, in their interrelatedness, and the belief that the earth is alive, not just inert matter to be exploited by techniques which are industrial in origin. The soil produces plants, which feed animals and humans, and if the various wastes are carefully returned to the soil as compost and manure its health and fertility are maintained or increased. Organic farming is, of necessity, mixed farming and makes use of rotational methods, whereas industrial approaches tend to be monocultural, bringing in their train featureless landscapes and the dangers of soil exhaustion. The attempt to obliterate pests by widespread application of chemical pesticides, another aspect of agribusiness, endangers variety by killing species of birds, insects and wild flowers; but the organic farmer works with a belief in a balance of nature provided by ecosystems. A certain level of pest and disease organisms can be tolerated where there is a variety of natural habitats providing cover and food for insects which are predators on aphids, so there should be no conflict between organic farming and environmental protection.[1] Pests will be less of a threat if plants and crops are healthy; plants and crops will be healthy if the soil is healthy.

But the interconnections extend beyond the farm and its landscape.

As Barry Wookey puts it, the philosophy behind organic farming, 'as with all things in nature, becomes more complex the more you think about it'.[2] He quotes the stated aim of the Soil Association, ' "to promote a fuller understanding of the vital relationship between the soil, plant, animal and man". (The Association) believes that these are part of one whole, and that nutrition derived from a balanced living soil is the greatest single contribution to health (wholeness)'.[3] The establishment of the Soil Association stemmed chiefly from concern for human health, its relation to diet, and the relation of nutritional quality to food production practices. The commitment of Lady Eve Balfour, its founder, was social as much as agricultural, and in many respects can be seen as typical of the period which produced the Beveridge Report.[4]

The organic outlook rejects any disjunction between social and agricultural issues. During the 1940s, other supporters of organic husbandry were arguing that the changes taking place in agriculture were bad not only for the health of individuals, but for the health of society at large.[5] A nation in which agriculture was no longer a central activity would become rootless and fragmented following the disappearance of traditional patterns related to natural cycles. Husbandry and craftsmanship were being replaced by the industrial mentality of exploitation and the profit motive; rural life was declining, society was becoming unbalanced, and an ever-increasing urban population was dependent for its food on a dwindling number of farmworkers.

Moreover, the effects of these changes had ramifications beyond the shores of this country; Britain itself had to be seen as part of a larger whole, linked with other countries and having a harmful effect on them through its policy of taking their food in exchange for industrial goods.[6] The industrial nations' attitude to natural resources is, from the organic outlook, a matter of global significance.

The final point to note is that support for organic methods of agriculture is often found in conjunction with a mystical or religious perception of the world: a sense of co-operating with a divinely-ordained pattern, to be understood rather than dominated or re-cast.[7] The material world is thus seen as inseparable from the spiritual.

* * *

'Monoculture attracts pests. Diversity helps to create a balance in the garden... We're not trying to wipe out pests anyway. We are aiming at a population of pests and their natural enemies at a level where the damage will not be appreciable.'[8]

These words of *Alan Gear* demonstrate that the organic outlook can apply to the smallest plot of land—the individual's garden. As he points out in his contribution to this symposium, the methods of 18 million British gardeners have a collective environmental impact which is unlikely to be negligible if nitrates and pesticides are used generally and habitually. If the greater organism of the natural world is to be healthy, it must consist of healthy cells and all who are gardeners can play their part by trying to work with nature rather than against it. This is not a negative matter, of simply avoiding chemical methods; it is a question of creating a healthy soil, employing crop rotations, and encouraging an ecological balance which involves animals, insects and birds. The effects of organic methods are then felt not just in the natural world, but in the human as well. Alan Gear begins his essay by referring to health issues and reminding us that a biological approach to cultivation is the only way to guarantee the beneficial properties of fresh fruit and vegetables. The successful organic garden is thus a microcosm of variety, balance and interconnectedness.

As the article by *Will Best* makes clear, the organic farmer has the same concerns, writ much larger. The opposition between the mechanistic and organic philosophies is exemplified on the one hand by large-scale, specialized, industrialized farming, with its tendency to monotonous landscapes and factory-line treatment of animals; and on the other hand by the mixed farm with its rotational variety, which avoids the risk of nitrate pollution, and makes use of waste materials by adhering to the rule of return—part of the cyclical, balanced pattern that is fundamental to the organic outlook. Like an organic garden, an organic farm will be aesthetically attractive and encourage wildlife. It is ironic that organic gardening and farming are often regarded as unsophisticated; perhaps the real, underlying objection is that they are more difficult than chemical methods, since they have to deal with a greater variety of factors and are necessarily more subtle and complex. As Will Best points out, the farmer who wishes to convert to organic methods cannot easily depart from a system dominated by chemical companies. The agricultural is inseparable from the economic and political.

Sir Richard Body, MP has been an outspoken critic of the influence of chemical companies on British agriculture, and of the Common Agricultural Policy (CAP). Here, though, he looks at the future of organic farming not from the perspective of complex bureaucratic regulations, but from that of the part-time farmer—the '21st-century peasant'. Perhaps this is how organic farming will establish itself within the interstices of European legislation. At present part-time farming is more significant on the continent than it is in Britain, but Sir Richard Body foresees its increase in this country. The idea of a hi-tech peasant need not be a contradiction in terms if 'peasant' is understood to mean somebody who takes a personal interest in his land and animals, cherishing rather than exploiting them. Sir Richard draws particular attention to the treatment of animals, which for the organic farmer will not be as mere machines for maximum production of food and milk. He suggests, too, that the part-time farmers themselves will benefit from living a balanced life—office-based and land-based, contemporary and traditional.

For *Michael Allaby*, the objection to pesticides is not that they are scientific, but that they are not scientific enough. A different concept of agricultural science is required—one that will be based on 'field ecology', to provide a more detailed knowledge of soil, insect, bird, animal and plant populations. Indiscriminate spraying of crops ignores organic interconnectedness, killing off creatures that are not targets, and asking no questions about longer-term results. What happens if there is a break in the chain of being? What happens if the effects of pesticides work their way up the chain? Loss of variety in nature follows and other problems, hitherto unimagined, are likely to develop. The organic outlook, conceiving of nature as a living web, is better able to face the complexities and will encourage a form of agriculture more respectful of, and responsive to, the patterns of the farming environment. In order to understand these patterns, argues Michael Allaby, agricultural science will have to develop increasingly sophisticated skills.

The discussion of the relationship between farming and the environment is developed in the essay by *Fiona Reynolds.* Intensive methods of cultivation have resulted in the loss of wildlife habitats and have produced increasingly featureless rural landscapes. The underlying assumption of this approach to farming is a mechanistic belief that animals and the soil are sources of productivity analogous to industrial plant, with efficiency, inadequately measured as output per worker, the sole criterion. Aesthetic considerations become irrelevant. Yet it may be that the attractiveness and variety of older landscape features have their utilitarian value too, as has been argued by Marion Shoard, for example.[9] Farming needs to become more sensitive to the benefits of a varied landscape, but Fiona Reynolds stresses that such sensitivity will require financial encouragement. British farmers are part of a larger whole, the Common Agricultural Policy, and any changes in their approach will have to occur within that wider political and economic context.

One particularly important aspect of environmental policy is dealt with by *Penny Evans.* The management of forestry and woodland provides another example of the way in which interrelationships are ignored. It has become separated from agriculture and treated as an industry in its own right, with monocultural swathes of forest covering large areas of landscape which are converted, in effect, to timber factories. Variety is lost, genetic stocks become limited, wildlife disappears, the soil is depleted, and the countryside used for the plantations grows drab and gloomy. No longer is there a link between trees and the requirements of farming, which now prefers to deal with metal rather than wood. Nevertheless, Penny Evans believes that an organic future for woodland and forestry policy is feasible, as there is an increasing recognition of the value of trees for a wide range of purposes, relating to wildlife, recreation, soil and water management, and atmospheric conditions. The mechanistic, mass-production approach must be replaced by a variety of methods appropriate to differing environmental surroundings.

Permaculture is one form of organic farming that attaches particular importance to the integration of trees as part of a varied system. It is chiefly associated with its tireless Australian advocate and practitioner Bill Mollison, but, as *Graham Bell* demonstrates in his article, its principles are applicable to the northern hemisphere just as to the southern. Permaculture exemplifies the features of variety, interconnectedness and a sense of the living whole and it implies, by its concern to observe and learn from what nature does, a reverence for the natural world rather than a desire to dominate. True, permaculture is a human activity which shapes nature, but the shaping is a creative partnership, a work of art that achieves its results by respecting the limitations of its materials and the laws which govern their interrelations.

Whatever social ideals may be cherished by supporters of the organic movement, the fact remains that changes in policy have to be brought about within the existing situation. Creating appropriate legislation is the politician's task. *Ron Davies, MP* argues that encouragement of organic farming is now possible, since the assumption that industrial methods could be sustained indefinitely is steadily being discredited. Public concern about the environmental costs of agribusiness is another significant factor in the change of public attitudes and widespread dissatisfaction with the CAP provides the international context. Ron Davies believes that organic farming, with its emphasis on diversity, rotation and good husbandry, will be beneficial agriculturally, aesthetically and economically. Even the National Farmers' Union is manifesting a degree of sympathy towards it—an unmistakable sign that the flaws in agribusiness methods are being revealed.

Turning to what is happening in other parts of the world, we find in the essay by *Wangari Maathai*, on the work of Africa's Green Belt Movement, a good example of the way in which an organic perspective can link spiritual belief with the material, social and economic necessities of existence. For Professor Maathai, trees are a gift of God which humans have wantonly destroyed, the price of such, almost blasphemous, thoughtlessness being the loss of the means of life. Destruction of trees upsets a region's ecological balance and, once that balance has disappeared, other losses follow. Food, fuel, resources, birds, insects and other animals are threatened, the danger of drought increases, and social fabrics disintegrate. The planting of trees leads to the creation of new cells which can develop into an ever more varied and healthy environmental, and therefore social, organism. The benefits of agroforestry need to be rediscovered, by Africans opposing their traditional wisdom to modern farming methods. Trees and shrubs aid the farmer by preventing soil erosion and it is within the power of everyone to plant them.

Organic farming is sometimes criticized for being backward-looking and unscientific, as the contemptuous phrase 'all muck and magic' implies. It is therefore something of a curiosity that the Japanese exponent of 'natural farming', Masanobu Fukuoka, rejects organic methods precisely because they are too similar to scientific agriculture.[10] The article by *John P. Reganold, Robert I. Papendick* and *James F. Parr*, reprinted from the June 1990 issue of *Scientific American*, might serve to confirm Fukuoka's fears, while providing a riposte to those who see organic farming as scientifically unjustifiable. The results of United States government surveys have suggested that organic farming is a highly effective way of sustaining a healthy soil, by combining conservation with the use of up-to-date techniques. As we would expect, the emphasis is on rotation, diversity, and a reliance on the balance of ecosystems for pest control, and the authors see the state of agriculture as inseparable from other issues, such as energy-use, pollution, and the economics of farming. Sustainable methods can bring benefits in all these areas and more. It is worth noting, too, that a recent study by the National Research Council in the United States confirms the views of Reganold, Papendick and Parr, and suggests that organic farms, if well run, can be just as productive as those using chemical methods.[11]

Whereas the authors of the *Scientific American* article concentrate on the practice of sustainable farming, *Wes Jackson*, a soil scientist, considers its social implications. His concern is for an approach to the land which will do justice both to the workings of natural systems, seen in a long-term perspective, and to the features of any given locality. Respect for the whole, and variety in the parts, create what he calls 'a mosaic of human cultures or communities'. He puts before us the contrasting images of two economic ideologies. One, representing consumerism, is linear, akin to the digestive tract. The organic alternative is a circulatory system. In Wes Jackson's view, the former is destructive because it does not respect nature's economy. A kind of 'dynamic accounting' is needed, which will look at how our treatment of the environment affects communities, rather than limiting its attention to numerical abstractions.

The wider concept of the organic embraces spiritual and political considerations; the study of developments in Colombia, written by *Peter Bunyard*, is a good example of this. What little hope there is for an organic future in South America requires an approach to the forests which is dependent on a religious reverence, but can be encouraged only by government action. Peter Bunyard therefore describes the changes in landholding laws that are gradually enabling Colombia's indigenous peoples to farm the forest land again, and the philosophy to which they adhere. The capitalist mentality is foreign to them and their economy is based on principles of exchange and reciprocity. Yet the survival of their harmonious approach to the forests' resources requires the support of the modern state. Such are the exigencies of interdependence.

The antagonism between traditional sustainability and the demands of commercialism is exemplified also in India. Capitalist propaganda emphasises consumer choice, but *Vandana Shiva* demonstrates in her essay that the consequent exploitation of resources reduces the variety offered by local peasant economies. She shows how, by regarding the Indian forest solely as the source of marketable timber, the species variety, which brings no financial reward, is diminished. Mechanistic knowledge is limited and utilitarian. The local peoples enjoy an extensive and systematic knowledge of the forest's diversity and know that the security of their agriculture and, therefore, of their societies, depends upon maintenance of that diversity. The reductionist paradigm of knowledge has to be challenged, as a matter not just of theory but of survival, and Vandana Shiva tells how groups of women, in particular, are resisting the powerful forces tending towards fragmentation and uniformity in their environment.

Of all the countries in the East, Japan has most completely adopted the techniques of western science and commercialism, so it is no surprise to find *Koyu Furusawa* writing about the possible extinction of a 2000-year-old culture based on the farming of rice. Japan is also well advanced in the process of finding alternatives which will be of interest for the organic movement in the industrialized nations of the West. After about only 30 years of mechanized agriculture, the traditional patterns of sustainable farming are being studied again to see what lessons they have to offer and the influence of co-operative movements, integrating producers and consumers, is spreading. In his history and description of these movements, Koyu Furusawa shows how they are typically 'organic' in the wider sense of their concern for the environment, health, and the balance of society. They bring the urban and the rural into contact with each other and help to restore relationships among neighbours, as the buying and preparation of food become communal activities. Koyu Furusawa also draws our attention to both the spiritual aspects of the organic philosophy and to the possible emergence of eco-fascism. His own emphasis is on an approach which avoids the extremes either of attempting to return to the Garden of Eden, like the advocate of natural farming, Masanobu Fukuoka, or of imposing a totalitarian solution. The organic co-operative movements are part of a pattern which includes economic, social and cultural realities, and will increase in strength by aiming at specific targets related to the variety of local needs and conditions.

Nevertheless, the question still remains whether a whole society based on a sustainable approach is achievable or even thinkable and, in the first of two essays, *Robert Waller* considers the general principles involved in any discussion of this far-reaching topic. A balance between past and present, tradition and progress, has somehow to be created. Like Furusawa, Robert Waller rejects the temptation to abandon all the creations of fallen humanity, but he insists on the importance of agriculture for any human society, a reminder of our dependence on the natural world. The Romantics protested against the mechanistic separation of the observer from the observed and its assumption of complete objectivity. Robert Waller argues that before a sustainable society can be created there has to be a shift to a philosophical viewpoint which takes account of the relationship between human beings and the natural world, despite their capacity to separate themselves from it intellectually. Human Ecology is such a viewpoint, and one which is organizing itself world-wide and fighting many local battles against the destructive application of the detached, mechanistic outlook.

The traditional patterns of agriculture and forestry could not have been replaced by industrial methods had power not been provided from non-renewable resources, and any discussion of an organic future has to consider alternative methods of energy provision. *Diana Schumacher* undertakes this task, describing both the general principles that must be observed and practical examples of their application. An organic energy policy will adopt a variety of approaches suited to the contexts in which energy is required, minimize harmful impacts on the environment, and seek to encourage the use of energy in forms related to cyclical patterns, rather than those involving a linear process of extraction, consumption, and polluting waste. The question of energy use illustrates very clearly the interrelatedness that is central to an organic outlook through an awareness of the dangers inherent in dependence on a particular, non-renewable fuel. Diana Schumacher's essay suggests that there are other, workable approaches which are more beneficial to individuals, societies and the natural world.

The cumulative results of our energy policy have become a matter of concern and debate, in so far as they may affect our planet's atmospheric conditions. *Professor Richard Grantham* discusses the steps which can be taken to reduce the so-called Greenhouse Effect, and outlines the approaches required for a policy of 'Geotherapy'. The implied metaphor is of Earth as an organism which has become sick owing to an imbalance in its system, and readers will be able to see the 'organic' connections between methods of agriculture, forms of energy use, squandering of natural resources, loss of species diversity, and the threat to the health of the biosphere. Professor Grantham develops the metaphor by suggesting that if the Earth and its biosphere can be seen as an organism, we would expect some form of evolution to occur. If human beings recognize their relatedness and responsibility to the planet they can enable the process of evolution to advance. The political dimension cannot be ignored, however. Various initiatives are being taken in different parts of the world, yet the problems are global and some sort of world authority or co-ordinating organization is required to give coherence and direction to the healing process. What will be the relationship between the whole and the parts? Professor Grantham concludes his essay by outlining a possible approach to the difficulty of combining diversity with integration in achieving solutions to world-wide problems.

In my own contribution, I examine the idea of the organic from a philosophical and historical perspective. We do no service to the organic movement by ignoring that the concept of the organic has a long and significant philosophical pedigree. If organic methods of farming and gardening are being reasserted against chemical and mechanized approaches, that is, at least in part, because an organic view of the world is believed to be more adequate and comprehensive than a mechanistic one. Organic farming emerged as a *conscious* opposition to industrialized methods of agriculture within a context of thought which interpreted the world as an organism rather than a machine. Through the history of this world-view there appear as constant themes the main ideas of the primacy of the whole, variety, interconnectedness, and aesthetic balance. Particular forms of organic philosophy, like vitalism, may be discredited, but the general features remain. However, the essay also contains a note of caution on relating the organic idea to social and political matters. The metaphor of an organism may be inadequate when applied to these areas, tending towards a dangerously totalitarian attitude.

The final essay, and the second contribution by Robert Waller, deals specifically with the religious and spiritual dimension to the organic philosophy—not in any abstract fashion, but by returning us to the earthy and elemental setting of the Gospels, and reminding us that the image of the Nativity symbolizes the relationship between the human spirit and consciousness and the natural world. Although there has been a good deal of hostility towards the Judaeo-Christian tradition in certain sections of the Green movement,[12] Robert Waller argues that the Gospels provide a spiritual teaching rooted in the peasant environment of a divine order. It is not, therefore, essential to attempt to transplant elements of other religious traditions into western culture in order to provide a spiritual basis for the organic movement. Instead, we should re-examine the origins of our own tradition and find within them truths that are still of value as we seek to preserve the spectrum of habitats, species and human capacities necessary for an organic future.

The essays which follow tend to emphasise the political, economic and cultural constraints which provide the context for any progress towards an organic future. As a result, the general feeling of the book may seem somewhat cautious. It might be suggested that political events in Europe during the last four years encourage the hope of widespread changes occurring with a rapidity that justifies optimism; or it might be argued that if, as a recent publication suggests,[13] we have only 15 years or so before the planet becomes irredeemably ruined, the tone of this symposium should be more urgent, more apocalyptic.

A sense of apocalypse can turn to a feeling of fatalism. It is the obverse of a desire for utopia, both extremes tending to discourage politically realistic action. So readers of this book will not find among the essays a vision of a dramatically transformed planet. Instead they will be given many examples of the way in which individuals, groups and institutions are doing what they can to move towards a world where the principles and practice of the organic philosophy are adopted. The fact that these changes are actually taking place should serve as the best form of encouragement that further advances can be consolidated.

Philip Conford
Chichester
January 1992

Chapter One

A Productive Partnership—Gardening with Nature

Alan Gear

HEALTH EXPERTS ARE INCREASINGLY COMING ROUND to the view that fresh fruit and vegetables are vital to the maintenance of good health. There is a corresponding concern about the potential long-term risks to health from consuming food containing pesticide residues—the result of modern farm practices. The answer lies in having organic food available at a reasonable price, but for the foreseeable future it is likely not only that demand will outstrip supply but also that the cost will be prohibitive to large numbers of people. The obvious answer, providing that you have a garden, is to grow it yourself. Not only can you be sure that it is truly organic, but you will have the advantage of the freshness that comes with home-grown produce.

The pesticide industry has its back to the wall, endlessly trying to persuade consumers that its products are safe, yet is undermined by one frightening report after another. The public is becoming increasingly nervous about the long-term effects of certain chemicals on human health. This lack of confidence in the products of modern agriculture extends throughout the industry. Worries about nitrates from fertilizers polluting water supplies, and flat disbelief about the safety claims regarding synthetic growth hormones injected into dairy cows, are just two examples of the many that exist.

Consumers are voting with their purses and switching to organic food, despite the fact that, at present, it costs more. If you cannot buy organic produce, either for reasons of availability or cost, the answer is to *grow your own.*

Organic gardening—common sense for the future

From being the province of a handful of 'eccentrics', the number of people practising organic gardening rocketed through the 1980s. Spurred on by the television series *All Muck and Magic?,* with a regular viewing audience of more than three million, the advance of organic gardening seems unstoppable. These days you would be hard pressed to find a TV or radio gardener prepared to defend the use of chemicals, other than in half-hearted or apologetic terms—'... if all else fails, spray with...' Hardly a ringing endorsement!

This is not to say that all the arguments have been won. Far from it. There is still an enormous educational job to be done. For example, far too many people regard organic gardening as simply 'doing without chemicals'. Throwing away your chemical arsenal is one thing, but just sitting back and leaving it all to nature is another. It is a recipe for disaster. Media reports of experiments in going organic, where the plot became overgrown with weeds and slugs ate all the lettuces, are far too common. Amusing though these reports may be, they still give the impression that the price to be paid for chemical-free growing is poor yields of crops that are devastated by pests and diseases.

Is organic gardening just a passing fad? Is there a risk of a backlash and of a subsequent return to the bad old chemical days? There are several reasons why I don't think this will happen.

We are all becoming much more aware of the overall impact of human activity on the environment. When we burn our garden rubbish we are contributing to atmospheric pollution and global warming, in exactly the same way as a farmer who burns the straw on his fields. When we liberally scatter fertilizer granules around the garden, there is every possibility that excess nitrate will make its way underground to join the excess fertilizer run-off from farm land. Wildlife is poisoned by pesticides sprayed on to crops by the gardener who uses fungicide on his roses or who sprays greenfly on his broad beans, just as wildlife is destroyed by modern farm practices. True, the scale is incomparably smaller, but consider that there are 18 million gardeners in the UK covering an area of a million acres. Collectively, what we do in the garden has an enormous impact, for good or otherwise.

There will continue to be a steady flow of stories about the harmful effects of agrochemicals and these will harden public attitudes against their use. For instance, dichlorvos is a pesticide with a range of uses. Mushroom growers spray it on to crops to kill flies in their mushroom houses. Nurserymen also use it to kill insects infesting their greenhouses. Until recently most people had never heard of it. All that changed when samples of sea-farmed salmon taken from supermarkets were found to contain excessive amounts of the chemical, which is also used to control sea lice. Immediately, consumer groups called for a ban on the use of this toxic and persistent chemical. If this is successful, dichlorvos will join DDT, aldrin, dieldrin, ioxynil and all the other chemicals that have been withdrawn from use following the discovery of undesirable side effects.

We have yet to see, for example, the full effects of high nitrate levels in drinking water. If, as some believe, it takes up to 40 years for fertilizer run-off to trickle down through the soil and rock before it reaches the water table, it bodes ill for the existing supplies that currently exceed EC nitrate limits. If they have got into this state as a result of fertilizer applied in the 1950s, how much worse will they become when the higher fertilizer inputs used in the 1960s and 1970s arrive.

Much more careful monitoring of pesticide residues, and scrutinizing of these results by independent bodies, is almost certain to throw up more examples of the dichlorvos type. Pesticides will continue to produce unwanted effects that could not have been predicted from the original safety testing. As a result, the public will demand ever more stringent tests which in turn will force up the cost of developing new products. Add to this the effect of higher oil prices on manufacturing costs, feeding through to the resulting price of chemicals in the shops, and suddenly the economics of chemical use becomes shaky.

The reasons cited above are bound to affect attitudes to chemicals overall. When customers go to a garden centre will they find themselves consciously avoiding the gardening equivalents of those products used by the farmers?

The immediate task of supporters of organic growing is to show that it is not only possible but essential that we grow crops without using chemicals.

Pest control without pesticides

Organic pest control relies on a multi-pronged approach rather than the simplistic 'if you see a pest, spray it' philosophy.

Preventive methods are all-important. This means choosing plants that are appropriate to your own soil, locality and the particular circumstances of your garden. Pests can be kept off crops by using physical barriers and traps, whilst a knowledge of the life-cycle of a particular insect pest is helpful in planning suitable times to sow to avoid trouble. Crop rotations, whereby the same broad families of crops are moved to different beds each year, help to prevent the build-up of diseases. Plant varieties which have in-built resistance to pests and diseases are chosen in preference to susceptible varieties.

You can also enlist the help of natural enemies in the garden. Beneficial insects are encouraged by growing plants which attract them. Other garden friends, such as hedgehogs and frogs, are made welcome by providing overwintering hibernation sites and ponds. Birds and bats, excellent aerial pest controllers, are attracted by nesting and roosting boxes and the planting of trees and shrubs, to provide shelter and food.

A couple of examples will illustrate how safe pest control works in practice. The carrot root fly is one of the worst carrot pests. The tell-tale signs of reddening of the foliage indicate that carrot root fly maggots are at work under the soil tunnelling into the carrots. Yet there is a simple and effective way of preventing damage. It has been found that carrot root flies fly low over the soil surface, rarely straying more than 12 inches above the ground. A two-foot-high vertical barrier placed around the crop keeps out most of the pests and damage is reduced to a minimum.

If a large quantity of carrots is being grown, or a windy, exposed garden precludes the use of barriers, an alternative tactic is to time sowings. Carrot flies usually go through two life-cycles in a growing season. The trick is to plan your sowing when the adult root flies are 'between' generations. In my own area, for instance, this means sowing later than usual, round about the first week of June. This is after the adult stage of the first generation has hatched and has already laid its eggs near to spring-sown carrots. They can then be lifted before the second generation is able to do much damage.

Slugs are one of the worst problems gardeners have to face. Chemical gardeners put down slug pellets containing metaldehyde or methiocarb but there is a good deal of unease about the potential effects of these chemicals on birds, hedgehogs and domestic pets. It is not even a permanent solution, as slugs will simply move in from surrounding gardens to take the place of those that have been poisoned. Regrettably, there is no one foolproof alternative method which organic gardeners can use to keep slugs under control. All we can hope to do is to reduce their numbers locally and protect plants when they are at a vulnerable stage. Large, clear, plastic bottles, of the sort that hold mineral water, with their bottoms cut off make excellent slug guards when placed over young plants. Similarly, barriers made from abrasive materials—soot, eggshells, sawdust, lime, etc—placed around susceptible plants will act as a deterrent to slugs unwilling to pass through. This recycling of materials is typical of organic gardeners. The problem with this latter solution, however, it that heavy rain will wash away the barriers—though they can be extremely effective in polythene tunnels and greenhouses.

Another tactic is to put out traps, in the form of dishes containing milk or beer. The slugs are attracted by the liquid, climb in for a drink and then drown. Some recent American research has shown that slugs have a definite preference for beer, so it is worth persevering to discover your slugs' favourite tipple.

Beer traps also attract ground beetles—these black, fearsome-looking creatures are greatly to be encouraged in gardens, because they are predatory on slugs and a number of other ground-living pests. Hedgehogs and frogs are also great slug controllers and, if encouraged through the provision of appropriate habitat, will help to keep numbers under control.

An area which is beginning to attract research attention is that of insect predators and parasites. Here, it is a question of trying to tip the balance in favour of natural enemies. The survival of ladybirds, newly emerging from hibernation, will be considerably enhanced if you grow a few nettle plants. The nettles are attacked by nettle aphids, which appear early in the spring and can provide a welcome snack for hungry ladybirds. Flat and open flowers such as the poached egg plant (*Limnanthes*), pot marigold (*Calendula*), and baby blue eyes (*Nemophila*) are attractive to hoverflies, whose larvae feed on unwanted greenflies.

Ladybirds, and to a lesser extent hoverflies, are 'high profile' beneficial insects in the garden. Of significant importance, but often overlooked, are the activities of parasitic wasps. These tiny creatures lay eggs inside the bodies of aphids and caterpillars, their presence detectable only by the minute hole through which they escape when adult. The adult wasps are attracted to plants of the umbellifer family (fennel, dill, carrots and parsnips) and by allowing these plants to flower in your garden you will do much to encourage them to stay. Recent research in the USA has shown that it is possible to train these wasps to be much more effective in finding their hosts. Newly emerged wasps are brought into contact with the scent of the insects which they will eventually parasitize. When they are released into the field they home in on the pests much more quickly than those which do not undergo this early learning process. The Americans are working towards mass release of such trained parasites as an alternative to spraying with chemicals.

The Henry Doubleday Research Association (HDRA) is active in this field by sponsoring research into the breeding of beetles for release as part of a slug control programme. The beetle in question is an indigenous species known as *Abax parallelopipedus.* It is found in woodland and so far is proving promising in laboratory experiments.

The release of organisms into the environment as a way of controlling pests is known as biological control. It has been little used in the UK, though there have been some notable successes overseas. To date most of the work in the UK has been concerned with commercial glasshouse growing. A pin-head-sized parasitic wasp has proved extremely effective in the control of glasshouse whitefly, whilst a predatory red spider mite effects control of its vegetarian cousin. Both creatures are now being used not only by organic growers but by a large number of conventional growers as well. These biological control methods, which used to be supplied only to commercial growers, have lately become increasingly available to amateur gardeners.

Over the last few years, research in this sphere has mushroomed, partly in response to consumer demand for crops grown with reduced pesticide input and also because more and more insect pests are becoming resistant to sprays. There are now biological controls for mealybugs, scale insects, certain fungal diseases and, most recently, for vine weevil. This latter pest had proved almost impossible to control by chemicals, to which it had become completely resistant.

The biological control consists of a predatory eelworm, or nematode, appropriately termed 'Nemesis'.

Looking ahead 25 years, it is not in the least fanciful to imagine garden centres resembling pet shops with an array of predators replacing today's chemical sprays. It is unlikely that we shall ever be able to do entirely without sprays but these would be looked on as a last resort using products derived from natural substances such as derris, pyrethrum or other low-toxicity, non-persistent compounds. Gardeners in the next century will look back in disbelief at the crude methods of our own time whereby indiscriminate use of chemicals killed off friend and foe alike. For my own part, I look forward to the day when I shall be able to go into a garden centre and ask for a bottle of beetles instead of a packet of slug pellets!

Fertility without fertilizers

Parallel with the development of biological and other non-chemical methods of pest control, the future will see an increase in the number and range of organic fertilizers. The market is already beginning to reflect this emphasis and new products appear with frequent regularity.

At the core is the recognition that the foundation of good gardening is the maintenance of a healthy soil. This means supplying the soil with organic matter. The best way of doing this is to make compost. In so doing we are able to recycle back to the land kitchen waste and garden rubbish that would otherwise be burnt or dumped. Indeed, it is possible that pressure from local authorities, unable to cope with rising amounts of domestic rubbish, may insist on waste being dealt with this way. How long will it be, I wonder, before all new houses are required to provide a compost box as part of the standard fixtures and fittings?

Of course, we will not only be using compost made from our own garden rubbish. Increasingly, other plant residues and industrial wastes are being turned into products that will enhance the fertility of our gardens. An example is a product made from decomposing straw with sewage sludge to make a friable soil conditioner. It may also be possible to use such materials, suitably formulated, as a replacement for peat in seed and potting composts. Research under way at HDRA is currently investigating the possibility.

As well as providing nutrients for soil micro-organisms, composted materials have a valuable role in maintaining soil structure. This can be particularly important in times of drought, as sandy soils fortified with more absorbent organic matter are better able to retain water. If we are to experience increasingly dry summers as part of the process of global warming, then complementary techniques such as mulching (covering the soil surface with organic matter) will be a vital part of growing. We are also likely to see a lot more use of green cover crops grown especially to conserve nutrients which might otherwise be washed out of the soil. These so called green manures are grown by only a few gardeners at present, but it is likely that, as more knowledge about them is gained, they will be taken up on a much greater scale.

Discovering more about green manures is one of the key research objectives of the Henry Doubleday Research Association. With more than 18,000 gardening members, it is Europe's largest organic organization. At its trial grounds at Ryton-on-Dunsmore, near Coventry, new methods of poisonless pest control and ways of improving soil fertility without chemicals are developed and tested. These can be seen at Ryton Gardens, the adjacent 10 acres of demonstration gardens which are open to the public all year round.

Every year the thousands of people who visit Ryton Gardens, and who take up organic gardening in their own gardens, recognize that growing organically does not mean decimated crops and paltry yields. Already there is a sufficient body of information to enable anybody to grow successfully without using chemicals. After years of neglect and travel down the dead end of chemical growing at war with nature, there is a dawning realization amongst scientists of the productive partnership gained from working with nature. The fruits of this relationship are beginning to be seen in biological and other novel methods of pest control and in a greater choice of organic materials. The future for organic growing has never looked brighter.

(Further information about the work of the HDRA can be obtained by writing to HDRA, The National Centre for Organic Gardening, Ryton-on-Dunsmore, Coventry CV8 3LG.)

Chapter Two

In the Balance—The Organic Farmer's Future

Will Best

On the face of it, organic farming has a future far rosier than at any time since its principles were first abandoned. Unprecedented media attention is focused upon it; increasing numbers of the population wish to support it and consume its products, which the food trade is now anxious to market; a national certification scheme is in place; various colleges and Agricultural Training Boards are laying on relevant courses. Two years ago the Minister of Agriculture announced a pilot scheme in which a hundred farmers who wish to convert their farms from 'conventional' to organic will be helped to do so through the EEC Extensification Scheme; this has not materialized yet, though a similar scheme in Germany is flourishing. But in fact it may well be that the future of real organic farming is hanging rather precariously in the balance. On the one hand, we see the inheritors of the organic tradition striving to farm their land according to the principles and not just the rules of the Soil Association—and in some cases at least receiving, for the first time, decent financial rewards for doing so—and on the other, an increasing band of food industry managers who can smell a profit in getting involved. Between the two we see the Minister and the Chairman of the UK Register of Organic Food Suppliers—the new regulatory body—committing themselves to neither side but mediating between the two. The fear is that the marketing men, who understand but dimly the motivations of those both producing and consuming organic food, may gain the upper hand and that there will be attracted into organic farming farmers whose need or desire to improve their financial situation is greater than their understanding of the principles involved. This could lead to the self-defeating situation of poor farming failing to fulfil its objectives of soil improvement, environmental improvement and the production of healthy food. It must be understood that current organic farming practice does fall short of the ideal system proposed by those whose ideas led to the formation of the Soil Association. But at the same time, the Soil Association Standards, to which most organic farming conforms, are seen by many in farming, including some who claim knowledge of organics, as being hopelessly idealistic and impracticable. The very processes of specialization, intensification and mechanization which were deplored by the early proponents of organic farming, have proceeded relentlessly since the war, so that the concept of rotational husbandry can be quite hard for a modern farmer to grasp in its entirety, while such is the stranglehold that the chemical companies have over farming that, until we have experienced it, we find it difficult to believe that any crop will actually grow without liberal supplies of their products being applied to it.

It would, perhaps, be useful to look at the system practised here at Manor Farm as, in the absence of many better examples, it is sometimes taken as being representative of contemporary organic farming. The farm, now 260 acres of fairly hilly chalkland, had been farmed, not particularly well, by my family for 50 years, when my wife Pam and I decided to convert it to a fully organic system in the early 1980s. At that time we were milking 90 cows, rearing our dairy replacements, and growing 80 acres of cereals. (I had been told 12 years before, by my college lecturer,that my only future lay in running one large, specialized dairy unit. Somehow I had not believed him—or at least I could not face the monotony of it—but that would have been the normal thing to do.) We had two full-time employees. Had we been specialized, the farm would probably have been more profitable, as the larger turnover would have enabled us to invest in better buildings and equipment to save labour and make us more economically efficient. We would also have been awarded a far larger milk quota, which is very valuable. There is no doubt that the financial rewards have gone to the large specialists who can make the biggest profits in good times, and shout the loudest in bad: in fact, for many farming families, to expand, specialize and mechanize has been the key to survival and those who have tried to carry on in the old ways have largely dropped out of business. This is the relentless march of progress which has applied to agriculture as to all other businesses. The damage to the soil, the landscape and rural communities has been clear for all to see, but the people on the ground have felt powerless to arrest it.

From our point of view it was the writing of Dr E.F. Schumacher in particular, and the self-sufficiency movement led by John Seymour, which first made us question our farming methods and helped us to see them as polluting and wasteful. Then the development of EEC surpluses made us very uncomfortable: there seemed to be little point in rising at 5 a.m. and working all day to produce something that was apparently unwanted. We also began to have grave doubts about the medicines which were being prescribed for our family and producing more side-effects than cures. Pam discovered homoeopathy, used it on our children and went on to study it. The value of organic food became clear. We went to a meeting organized by the Soil Association and found there was a small nucleus of people deeply involved in organic farming and growing and determined that it had a future. Through British Organic Farmers and the Organic Growers Association these people were running seminars, farm walks and the biennial Cirencester Conference. We found out about the Elm Farm Research Centre, set up as a charitable trust in Berkshire to develop and spread knowledge of organic farming systems. Here they were beginning to do conversion plans and were pleased to help with ours.

The aim of a conversion plan is to devise a rotation which can be developed from the existing system. This rotation should be balanced for nutrients, alternating fertility-building with fertility-using crops, deep-rooters with shallow-rooters, allowing cultivation for weed control, and tending to break pest and disease cycles. Normally this involves livestock and grass/clover leys, but in some cases stockless rotations are being tried, using leguminous crops and green manures. Since our farm was already mixed to some extent, we were in a good position to put it on to a proper rotational basis and farm it organically. We now milk 70 cows and rear their replacements, lamb 100 ewes, grow 70 acres of cereals and keep some pigs. Of the cereals, 25 acres are wheat grown both for flour and for thatching reed and the rest are barley and oats, mostly fed to our own stock. The basis of our rotation is the four-year ley, grazed and cut, and fertilized with manure from grazing and yarded stock. The only major investment we made was the erection of a covered yard to replace our cow cubicles, so that we can bed our cows with plenty of straw and make decent manure rather than slurry, as well as keeping them in much greater comfort. We also sold a small field to a couple who grow a variety of vegetables, working in co-operation with us. We now have one full-time and three part-time staff. The rotation followed now is, over most of the farm: four-year ley/wheat/oats, two-year ley/wheat/barley, but in the fields close to the dairy: four-year ley/grazing rye/kale/oats. The four-year ley contains tetraploid and perennial ryegrasses, white clover, chicory, cocksfoot, timothy and other herbs, while the two-year ley comprises ryegrasses, red clover, large-leafed white clover and chicory. Silage-making and grazing with cattle and sheep are alternated to minimize worm problems in the livestock and utilize the swards most efficiently. We have also grown lucerne in the past; it is a good crop in an organic system, fixing much nitrogen and making excellent silage, but we had problems with docks where we were growing it as a pure stand. We will probably include it in mixtures in the future. There is also a small acreage of permanent pasture, a 10-acre wood and some scrub land. We manage most of our hedges rotationally as well so that there is always some freshly laid hedge and others in various stages of growth. The whole farm is, therefore, attractive to humans and to wildlife. This is something which organic farmers regard as important and is written in to the Soil Association Standards. Our aim always was to hold the Soil Association Symbol and in 1988, five years after starting the conversion, we were awarded it for the whole farm and all enterprises.

The level of mechanization on the farm would appal our forefathers and is open to criticism for the amount of fuel we burn. Although we do far more manual work than many conventional farmers, we still have three tractors: their principal use is in silage-making and muck-handling, in addition to cultivation. It is interesting to note that Newman Turner, describing his 180-acre organic farm in 1950, writes: 'I have some steep hills, and must therefore have a fairly high-powered tractor... a twenty-seven horse-power model.' Well, we have an 85 horse-power model and a 70 horse-power one, which a neighbour recently referred to as 'useful little four-wheel-drives'. I really cannot see how we could do without them—which probably shows a lack of imagination on my part—but what we can do is to try to minimize fuel use: use them as efficiently as possible and cut out unnecessary journeys. There is no reference in the Standards to this issue and at this stage it cannot be tackled seriously, but sometime in the future it will have to be addressed. We do walk a lot, and also have a mountain bike.

The other major departure we make from purity is the importation of some fertility. We use no chemicals, but we do use some products of chemical systems. The Standards allow the use of conventionally grown straw for bedding, and every year we bale 60 acres of wheat straw from a neighbour's farm, which would otherwise be burnt, and litter our cows up with it. Twenty per cent of the animal's dry matter intake may be from certain conventional sources, and we buy in feeds such as maize gluten and distillers' grains to balance the cereals in our cows' concentrate ration. I have two comments to make on this. The first is that as we proceed with our farming I hope that we will steadily build natural fertility and be able to move to a more balanced system with fewer cows and more wheat, which will reduce our need to import straw as well as producing a more balanced output in terms of human nutrition. The second is that if we are denied the use of the by-products of our production because they are flushed down sewers or tipped into land-fill sites, causing all sorts of pollution problems, then we may need to replenish those nutrients from some other source. This is another issue which will surely have to be addressed at some time if we are to have a future, but is quite out of the control of the ordinary farmer.

What of the economics of this farm? Over the last two years farming in general has suffered badly in the recession, and we have not escaped this, but our financial performance is no worse than it would be if we were not organic. Our physical output is about 70 per cent of that of a conventional farm, while our labour and management costs are considerably higher, but we have no bills for crop chemicals or normal medicinal drugs. The health of our crops depends on the health of our soil, while our animals keep fit grazing our organic leys. Illnesses are treated with homoeopathic medicine, itself cheaper than antibiotics. But the principal factor in our financial viability is the price we receive for our products. All our produce is now sold at a premium, quite simply because the market is seriously under-supplied. The whole concept of having to pay over the odds and not being able to dictate terms to their suppliers is quite alien to the multiple retailers who dominate the food trade, but at the moment, if they want organic food on their shelves, they have no alternative. The premia vary with the enterprise, but between them they increase our turnover enough to compensate for our reduced yields, as compared to normal farming.

Herein lies the central dilemma: premia are at once the salvation and the potential downfall of organic farming. Without them it would not be viable, but by the time they have been multiplied up the retail chain, the food becomes very expensive. Tell a farmer about them and he may consider converting his farm; mention them to a conservationist and he is likely to accuse you of doing it for the money and not for the planet. If prices are too low, then genuine organic farmers cannot survive; if they are too high, then unscrupulous profiteers may try to jump on the bandwagon. Anyone who is genuinely committed to an organic future must want to see all our land farmed properly and all our people able to eat organic food, but has to be aware that if this happy state of affairs were to be even remotely approached, the premia would disappear and some other means of support would have to be found. These premia are well liked by the Government and the economists. Organic farming may become self-supporting and not require any state funding: those bearded philosophers have found themselves a 'niche market' and become entrepreneurial capitalists. 'Organic farming can never feed the nation', says Mr Gummer, 'but it can play a serious role in supplying its own specialized market'. This way he can give some help to the organic movement whilst neatly avoiding all the serious problems caused by modern farming and not upsetting the powerful interests vested in the status quo. The chemical companies hold sway over most of the farming scene and some of the more obvious results, such as over-production and nitrate pollution, are tackled in a piecemeal way with such schemes as 'nitrogen-sensitive areas' and the rightly much-maligned 'set-aside'. It is easy to become frustrated by what seems to be the stubbornness of those in authority in refusing to acknowledge the extent of the problems and the way in which organic farming could provide the answers. But we have to remember that a well-run conventional farm produces previously undreamt-of yields of cheap food, while many organic crops have been, and still are, thin and weedy. Conventional farming can look very good, while the problems may be well hidden, particularly in this country with our moderate climate and relatively stable soils; organic farming may look a little ragged, while the benefits to the soil and the air and the groundwater and to those who consume the food are not immediately obvious.

I do believe that now is not the moment to worry unduly about all the possible pitfalls, but to grasp the opportunities which are presenting themselves to forward the cause of organic agriculture on all fronts. So research, education and training, media and consumer interest, the Government's extensification scheme which is intended to include some payments for organic conversion, and marketing schemes may all be of potential benefit. A farmer intending to convert his farm can now attend college courses and training days, have access to research and join a co-op to sell his produce; none of these was available a few years ago. What is vital is that these developments should be guided by people who really understand what organic farming is all about. The educators and researchers have to be educated, and the civil servants who administer support and certification have to understand what it is that they are supporting and certifying. If this can be done, then there is every chance that the organic farms which result will be good enough to make the wisdom of conversion widely acceptable. And if this happens, then the wider task of converting more than a few obviously suitable farms can begin.

It could be seen earlier that the conversion of my farm was straightforward, because it was already mixed and we had a dairy herd and a range of implements, buildings and skills. To convert a specialized arable farm would be far more difficult and expensive. Restoring humus levels and building up real fertility on land that has been continuously cropped for decades can be a long process. Many of these farms have none of the infrastructure needed to keep livestock, so buildings, fencing and water supplies would need to be laid on as well as grassland machinery obtained, in addition to the purchase of the stock and the recruitment of skilled labour. At the same time, much of the arable equipment would become under-utilized, but would still be needed. Organic management is more difficult than conventional anyway, and as soon as you introduce more enterprises you increase the complexity and multiply the paperwork. Similarly, a specialized dairy or livestock farmer faces equivalent problems when he decides to reduce his stocking and introduce some cropping, but I see no alternative if organic farming is to become widespread. At the moment there is a terrible pollution-causing waste of dung and urine on livestock farms and straw on arable farms. Separate the two and you have all sorts of problems; put them together and you have humus-building, life-enhancing farmyard manure. Given the great East–West divide between arable and livestock farming in this country, any movement towards an organic future is bound to be a very slow process but at least at the moment it does not seem impossible.What is essential is that the organic farms operating in the next decade must be good enough to convince the various interested parties of their economic viability, agricultural productivity, sustainability, ecological balance, and of the real value of the food they produce. This is a tall order, but, if we fail now, a great opportunity will have been missed.

Chapter Three

The 21st-Century Peasant

Richard Body

A VISIT TO THE WEST COUNTRY made me realize that a great transformation is certain to come to agriculture. I was there to look at a farm. The farm itself was of little interest; it was the farmer who set me thinking. He was a commodity broker in the City of London. Until a year or so previously, he could have done his job only in the heart of the City, on the floor of some exchange, no doubt tick-tacking madly at other dealers as prices went up or down. A hectic life, far from idyllic, the very opposite of anything bucolic.

Yet here he was in the midst of the countryside running a 200-acre farm while doing precisely the same commodity broking as he had been less than two years previously. A room in the farmhouse had been rigged up as his office with various bits of electronic gadgetry that enabled him to do his buying and selling as effectively as if he were back on the floor of the exchange. Once a week, sometimes less frequently, he travelled up to the City to meet various people and do the few things that could not be done quite so satisfactorily at home. Besides, if he were to stay away from the City all the time, all his nervous energy might seep away and the stimulation of talking to colleagues face to face would be missed.

Still, it made me think. If such a job as his could be done in a faraway farmhouse, how many others could also be done? Five million? Ten million? Estimates have been made by others and it seems probable that by the turn of the century several million men and women will be able to quit our cities and conurbations and continue their employment a hundred or more miles away. The number will increase until only a minority need remain in the towns.

The emigration may be checked by employers being reluctant to spend considerable sums of money on the gadgetry, while many employees will wish to remain among the concrete. On the other hand, better work may be achieved in a more congenial environment. No travelling by bus, car, underground or British Rail for an hour or more each morning will be a positive gain; being able to lie in bed an extra hour will tempt quite a few; and seeing more of one's spouse and children may or may not be a further benefit.

So it's anyone's guess how many will make the move. Let us assume it will be no more than two million—that must be a very modest estimate. How many of them would wish to shrug off all their suburban values and immerse themselves in country living, must also be in the realm of surmise. While only a few might have the inclination, let alone the capital, to take over a 200-acre farm, it might not be unreasonable to expect one in ten to dabble in some form of agriculture. That is 200,000, which is much the same as the number who are farming today.

It could well be many more than that figure. After all, once transformed into teleworkers, they will have plenty of spare time. The opportunity at least to grow their own vegetables and keep a few hens or a beehive will be available to a huge proportion of the population. Naturally, a majority may wish to spend most of their new leisure time on the golf course, at the tennis club or on some other leisure pursuit begun in their earlier life. Others will take up a serious hobby or write off to the Open University. But producing food—their own food—will have an appeal to a very large number, especially as it will be an antidote to the very exacting and unnatural work they will be doing on their machines. Being out-of-doors and constructive, as well as satisfying the basic instinct of self-providing, will be a total contrast to their ordinary work. What for them could be more recreational, in the true sense of the word? Also, if redundancy came, the family would not starve.

Perhaps the reader begins to share my vision. Many will be content to produce only for themselves, but the 200,000 I have in mind will be engaged in serious part-time farming and, like today's modern farmers, they will not be motivated by self-sufficiency but engaged in securing a supplementary income from the market.

The implications for agriculture are immense. As they will be taking to the land because they believe they will enjoy doing so, they will be farming as they think it ought to be done. Not primarily motivated by profit or the necessity to earn a livelihood out of their acres, they are obvious recruits to organic husbandry. I am not suggesting they will all be of one mind, but a decade hence that is likely to be their inclination. Organic farming is now coming into vogue. Many new entrants today suspect it has an inevitable future, so if that is the view in 1992, how much stronger will the feeling be 10 years from now? We may hear a great deal more about the dangers of nitrates and pesticides, hormones and antibiotics in the next few years, and these newcomers will have no wish to endanger their families with poisons from the chemical industry.

They will begin, most of them, with a healthy prejudice in favour of organic farming. What kind they will take up will depend largely on where they have chosen to settle. Many of the Londoners will make for East Anglia where the soil, climate and local tradition will introduce them to arable crops. What price, then, the prairies of the grain barons? In the nineties, the surpluses of wheat and barley (sadly of inferior quality in England, compared with what the Continentals produce with a more favourable climate) will continue to force down prices, certainly in real, if not nominal, terms, and as inflation is bound to continue at some rate, the present arable farmers will be caught in a cost-price squeeze, so that their higher costs will not be met with higher prices. As thousands of them claim that they cannot endure the existing squeeze much longer, their future is doubtful, particularly the ones with borrowed money.

So I see many of them selling their acres to the teleworkers. As none of the latter will be minded to buy 2,000-acre holdings, we should see these large farms being divided, which simply means they will be going back to the size of farms that existed when the exodus from the land began.

In East Anglia, many co-operatives exist already for the ownership of combines, pea harvesters and other items of expensive machinery that are used only a few days in the year on the farm. Such co-operatives would enable the incoming peasants to carry on with those arable crops which require costly equipment.

With their main income obtained from elsewhere, the necessity to make the farm pay will not be the paramount objective. A profit at the end of the year will be an index of success, something gratifying but not essential. A modest loss may be seen as a signal that other methods had better be tried, and so long as the year's work has brought some enjoyment, perhaps a sense of physical fitness or self-fulfilment, the intangible gain will be set against the tangible loss.

Hobby farmers have often made losses. Indeed, they usually do. They do not take the job as seriously as the working neighbour. Jobs on the farm have to take second place to the 'real one' and much has to be postponed to the weekend, although the delay will cost money. The hobby farmer will also experiment. He can afford the risk and his experience or skill acquired off the farm may give him ideas. Most of the great agriculturalists have taken up farming as a hobby. Jethro Tull, the greatest of them all, was a London barrister. Experimenting adds a little spice to any job. The risk of failure is a challenge and, when it materializes, a financial price has to be paid. Not unnaturally, the ordinary working farmer is loath to leap into the uncertain darkness.

So what will happen with 200,000 new hobby farmers? Farm-gate prices in real terms have been steadily falling ever since 1950 (except for the temporary increase when import levies were introduced) and the fall has forced from the land those unable or unwilling to increase their output either by taking over a nearby farm or pushing up the yield of their existing soil or stock. They have been on a treadmill. To stand still in terms of income, they have had to work harder or, more accurately, to get more out of their farm. Farmers have been falling off the treadmill at the rate of 3,000 or 4,000 a year: the 500,000 farmers in the UK of 1947 have become 200,000.

The danger to these survivors of a similar number of hobby farmers is pretty plain. The 21st-century peasant will not positively want farm-gate prices to go on falling, but if they do, violent demonstrations with hundreds of tractors blocking Whitehall and thousands of sheep turned loose in Piccadilly, are not very likely. Our new peasants will be at their electronic gadgetry a hundred miles away earning an income.

Many of the new peasantry will be delighted to see farm-gate prices come down. Wheat at £60 a ton instead of £110 will bring down the price of land to enable our peasant to buy a little more of it, as well as give his younger brother the chance to sell up in Surbiton. There is a danger of too many newcomers coming into the countryside at once and so putting up land prices too quickly. In one respect, this will be an advantage. The existing farmers will be able to sell at a reasonable price to compensate for their falling income. But the hordes will not come all of a sudden. The treadmill may accelerate with 10,000 working farmers falling off instead of less than half that number and, if most of their farms are divided in two, it would not take so many years for the 200,000 new peasants to be accommodated.

Would any full-time working farmers survive? Life would become very difficult. A nil rate of inflation by containing a rise in costs could, in the short term, be set against the fall in farm-gate prices. Yet who will predict a period of no inflation, the first since 1939? The steady exodus from the land since 1950 has so weakened the farming lobby, and the National Farmers' Union in particular, that already it is cut down to the size and venom of a paper tiger. The further loss of full-time working farmers will reduce to threads any organization capable of successfully demanding higher farm-gate prices. This makes the future of the conventional farmer bleak indeed.

The future of conventional farming itself may likewise become bleak. Let us see what is likely to happen to the livestock side. The probability is that the 21st-century peasant will wish to keep some stock whatever holding he has. Will he derive much joy in keeping 50,000 hens in batteries or 100 sows in stalls so narrow that for four months at a time, from when they are served by the boar to when they farrow, they cannot move, except stand up or lie down? Even if the sight sent him into ecstatic raptures of delight, one of his main concerns would be to eat food more palatable and fresher than what may be available on the supermarket shelf. This, after all, has been one of the motives of the 'good-lifers' and our new peasantry is not going to be such a different breed. Besides, the medication he pours into his intensively kept stock may make him wonder whether the profitability of his factory farm is scarcely worth the candle when the same food can be bought in the shop. We can be sure his hens will be running around either freely about his yard or rather less freely in some barn system; they will be healthier birds laying healthier eggs. If my own feeble taste buds can invariably tell whether or not an egg comes out of a battery, I am sure this new legion of peasants will also be capable of doing so. Why keep half a dozen hens for domestic egg consumption when five or six times as many are little extra trouble and surplus eggs can be sold? With a score or more of these holdings, each having surplus eggs for sale, it will be a useful sideline for someone to buy a van to start a collection for them to be taken to a centre where, with thousands of others, they are graded and packed for sale to the retail trade.

Much the same will happen to the pigs and cattle. More humane methods of husbandry will come back because the new peasants will have a set of moral values which make them recoil from cruelty. The modern conventional farmer has been driven to various forms of intensive methods, which in all probability his father or grandfather would have rejected, because the cost-price squeeze has allowed him no alternative. Some farmers have taken intensive methods several stages further than was necessary; they are the greedy ones, the kind who regard animals as no more than machines for making money, and of this type we can be sure the 21st-century peasant will not be. If he wants to make a lot of money, he will know it will be wiser for him to stick to his electronic gadgetry.

Organic meat has to be of certain standards, humane rearing, transportation and slaughter being among them. These will appeal, if not to all the new peasants, at least to a majority. A recreation that is not satisfying is not one to be pursued for long. As it will be more satisfying to keep animals humanely than otherwise, our farm animals will bless the day when the new peasants take over. The pigs will be out in the fields in summer. When they are likely to thrive better indoors, then in they will come and so, too, will the cattle.

As with the hens, to keep additional pigs or cattle will occasion not much extra work, and the surplus will be for the market. The five or six chains of supermarkets that dominate the retail trade in food will have to respond to this change of supply. The conventional farmer has produced at the behest of the supermarkets, for he has had no choice. Like all peasants throughout history, ours can be sanguinary-minded. The producer, at last, will not be dependent upon the dictation of the chief buyer and a different relationship between the food producer and the food retailer will emerge. If they are wise, the food retailers will adapt. As it is, we have several of them only too anxious to buy more organically grown food. It is in their interests to sell what is safe to eat and good tasting. That points in the direction of organic food. But much of it lacks eye appeal, the kind of quality that the consumer has preferred. Most of us in the business of food production know that this eye appeal can be obtained only by the use of pesticides, preservatives and colouring agents, none of which are positively good for us and at best are neutral in their effect. As our peasants will be loath to use these chemicals, I can see a sensible coalition being forced between them and the retail traders. The coalition will have a single message—to persuade the consumer to forget about eye appeal and go for taste, safety and nutrition. The quality of the food is internal, not external, will be the essence of the message.

Now we cannot speak of food safety without mentioning the great army of pesticides. As they are used in horticulture and on the arable side, our new peasants may play a notable part in changing this aspect of agriculture. By wanting to eat their own fresh, palatable, nutritious food, a natural prejudice towards organic husbandry will be found, almost inevitably. The next decade will hear more, not less, about the dangers of pesticides, for as the old pesticides lose their efficacy and have to be replaced by new and stronger ones, the public's existing concern about their effect upon human health is bound to intensify.

But pesticides also kill our wildlife. Will our peasant wish to give up the wildlifeless environment of the city to find it again in the countryside? If he is indifferent about whether the birds sing, the wild flowers blossom or the butterflies fly, then he is not the peasant of my dreams. A desire to safeguard the wildlife from the poisons of the chemical industry may not be his highest priority, but at least it will have a place in his order of things.

The 21st-century peasants are going to come. The wonders of modern technology make it certain; and while we may be unsure of the numbers or the speed at which they will come, we can be sure their influence upon farming will be benign and good. It follows, does it not, that with them organic husbandry has an immense future.

Chapter Four

Phasing Out Pesticides

Michael Allaby

HALF A CENTURY AGO, the people of Britain faced the possibility of real hunger. They did not know it at the time, of course. Such demoralizing news was kept secret. Had there been an effective blockade of all the sea approaches to this country, British farm output would have been insufficient to provide an adequate diet for everyone. We would have had to choose between capitulation and starvation.

Apart from a brief revival during the 1914–18 war, British farming had been deeply depressed ever since the opening of the North American railroads and the invention of refrigeration for ships had made it possible for us to import food more cheaply than we could produce it ourselves. In the end though, the 'cheap food policy', so beloved of market-oriented politicians, brought us to the brink of catastrophe.

All the changes made in farming since then, all the 'modernization', the 'industrialization', the 'chemicalization', spring from that experience. The strategic importance of agriculture had to be recognized and Britain had to aim for something approaching self-sufficiency. The policies that began in the 1940s were designed to provide economic stability for farmers by guaranteeing their incomes and thus to encourage investment in a seriously under-capitalized industry. The introduction of new techniques and tools was made possible and payments were adjusted to encourage a rapid increase in output, which was achieved mainly in a great burst of expansion during the 1950s and 60s.

The introduction of modern pesticides formed part of the change and contributed to the expansion. As the Soil Association was pointing out 30 years ago, however, it was almost impossible to tell how big a contribution they made. Everything happened together. Grassland was ploughed up and sown to cereals, producing our nitrate pollution problem many years later, fertilizer use increased, new machines made timely cultivations and rapid harvesting easier, and herbicides and insecticides were used. How can you separate their effects?

Our entry into the European Economic Community is usually represented as a means of gaining access to a market for industrial goods of some 200 million people. That may be so, but it also gave us access to European agricultural markets and products. Self-sufficiency became more easily attainable if the unit were the whole EEC and in fact it was soon achieved. As Europeans, we are now self-sufficient in all commodities that can be produced in our climates. Indeed, we have been rather too successful, for we now chronically over-produce certain items and complain at the drain on public funds of buying them for storage or disposal.

You might think that having attained a goal most people desired we would be happy, even smug, but we are not like that. We complain of the cost of the Common Agricultural Policy, at the waste implied by over-production and subsequent disposal, at the changes modern farming has wrought in the countryside, and we fear that some of the methods used may be unsafe. In particular, we worry about pesticides. It seems inherently dangerous to make widespread use on food products of substances that are, by definition, poisons.

Our attitude to the countryside and to the activities of farmers depends critically on the availability of food. This is nothing new. It has been so throughout history and is recorded in what commentators regarded as an aesthetically pleasing landscape. When food was scarce, the attractive countryside was the farmed countryside. Travellers liked to see neat fields growing healthy crops and the heaths, marshes and mountains were considered bleak, dark, hostile wildernesses best avoided. When food was plentiful, the wildernesses ceased to threaten, for unproductive land could be afforded; they became romantically appealing, and farmers were berated for attempting to plough or drain them.

Such changing attitudes must not be seen as frivolous. It is inevitable that our view of the world will be conditioned by our recent experience and our expectations and, therefore, that our view will change from time to time. In the years following the 1940s most people would probably have preferred to see a 'working' countryside with the promise it held of a secure supply of food. That may explain the long, hard struggle the Soil Association faced in its efforts to pursue ideas that were, literally, unpopular. Today preferences have changed, but this does not mean in any absolute sense that one set of ideas is true and the other false. Each is appropriate to its time, and is likely to change as times change.

We are not much different from our ancestors, except that we are much more mobile and instead of one prevailing idea of the kind of countryside we prefer we have three, and they conflict. Farmers favour neat, 'clean' fields where not a leaf is out of place and no pest or weed is to be seen. Naturalists enjoy a wilder scene in which human intervention is reduced to a minimum and is designed to sustain and improve natural habitats. Most visitors to the countryside hope to find attractive landscapes, open areas to which they are allowed access, and facilities at least for refreshment. They have little knowledge of farming or sympathy for 'modern' methods that detract from the idealized rural scene they seek, but they are by far the largest group. Informal recreation in the countryside is now much the most popular leisure activity and its popularity is increasing rapidly. More people see more countryside than ever before, and their demands cannot be ignored. So far as farmers are concerned, this social change presents both a challenge and an opportunity.

The challenge is obvious. People are concerned, 'green' issues are impossible to ignore politically, and if environmentalists attract widespread popular support for changes in farming, sooner or later farming will have to accommodate those changes. The opportunity arises from the possibility that such changes may be made financially possible and even attractive.

It seems inevitable that pesticide use will become much more strictly regulated and possibly legislation may reduce it substantially or even phase it out altogether. The question 'Is this possible?' hardly makes sense. If that is what is decided, that is what will happen; therefore it is necessarily possible.

The question 'Is it desirable?' is more apt. Clearly there is a problem, in perception and in reality. The problem has been somewhat exaggerated during its transfer from reality to perception. In these days, when opinion is formed and controlled by the press and television, in turn influenced, not to say manipulated, by well-organized pressure groups, distortion may be inevitable and as facts give way to emotion it is not so much truth that falls victim as our sense of proportion.

I should say, therefore, as a matter of fact, that the British population is not being poisoned by food contaminated with pesticides. Indeed, so far as I know, there is no case of a consumer being injured, far less severely injured or killed, as a consequence of eating food that has been sprayed. When chemical products are described as 'carcinogenic' it is on the basis of laboratory tests using tissue cultures or experimental animals, and it supposes, sensibly enough, a dose much larger than that to which any consumer is likely to be exposed. It may be that eventually some pesticides will turn out to have caused actual harm to human consumers, but there is no evidence of it at present.

The regulations governing the use of pesticides are generally observed, but this is not to say there are no accidents of any kind. Farmers can be careless, even criminally so, just like any other group of people. In 1988, 160 complaints about pesticide misuse were investigated by the authorities and in at least two cases crops were destroyed because growers were attempting to market them too soon after their final spraying. Agricultural workers have been harmed by handling pesticides, although such cases are not common. People have been made ill when drift from nearby sprayers has exposed them to pesticides directly, and garden crops have been damaged by drifting herbicides.

Pesticides have certainly harmed wildlife, mainly because of the concentration along food chains of the more persistent compounds. This problem is well known and species at the top of food chains, particularly birds of prey, have been recovering since the use of those compounds was drastically curtailed or banned.

It is hardly surprising that pesticides harm wildlife. That is what they are meant to do, if you allow that pests and weeds are also wildlife. There may be an irreducible minimum to such incidental harm, and the injuries are not new. Before the introduction of modern compounds some of the substances used must have injured species other than those at which they were aimed. Caustic soda mixed with lime was recommended for slugs and snails, for example, copper sulphate for some weeds including couch grass, and compounds of arsenic, used to kill both insects and weeds, included such alarming concoctions as lead arsenate and copper acetoarsenate.

The organochlorines that replaced these simple compounds were actually much safer. The toxicity of DDT to humans, for example, is exactly the same as that of aspirin, and although it is soluble in lipids and so is stored in body fat, it seems to cause no damage and in time it is excreted. The persistence of the organochlorines was one of their main attractions, for it reduced the number of applications that were needed, but in the end it sealed their fate because it allowed them to be concentrated ecologically. They gave way to organophosphates, which are less persistent but in some cases very much more toxic. Since the organophosphates were introduced, a whole range of compounds has followed.

New products are developed in order to provide more effective treatments and generally this means more precise and safer treatments. It is not easy, however, and we may be close to the end of this particular road. The cost of developing a new pesticide is huge, and once the research and development are completed and production begins, the manufacturer has no choice but to try to market the product as widely as possible. The high development cost arises partly from the extensive safety testing that is required. If testing requirements become more stringent, then development costs are bound to rise.

One of the most frequent complaints lodged against the early compounds is that they were non-specific. They killed too wide a range of non-target species, including species that were natural predators of the pest and so might have exerted a measure of biological control. In some cases this problem can be solved, at least in theory, although in the real world perhaps only in theory. A thorough knowledge of a pest may make it possible to design a substance that attacks that species and no other. Unfortunately, the resultant compound will be of use only against that particular pest and so the market for it will be small, but the development cost even higher than that for a more broadly-based poison. A safer product may be feasible technically but impossible economically. In other cases the technical problem is more difficult. Consider what is involved in devising a product that will kill some grass species but not the others among which they grow.

Tougher regulations, therefore, may actually force a reduction in pesticide use. It will become too expensive. The initial investment will be too large, the 'lead time' between the first design and a marketable product too long, and the eventual price unattractive to farmers. Manufacturers will be discouraged, investment will be reduced, and factories will turn to other work. So far as the chemical industry is concerned this may be inconvenient but it will not be disastrous. Most pesticide manufacturers already have other, non-agricultural sides to their businesses, and it would not be too difficult to find alternative employment for their laboratories, industrial plant and raw materials.

Such restrictions will result from consumer pressure. Farmers, chemists and toxicologists may consider such pressure unwarranted, but in my opinion it will be imperative. The risk of poisoning by pesticides will join fears of bacterial contamination of eggs and meat products and a general distaste for intensive methods of livestock husbandry on the grounds of animal welfare. At the same time, demand for improved access to farmed land and anger at the apparently arbitrary closure or diversion of rights of way will bring pressure from another quarter. The naturalists are exerting yet more pressure for land in what have already been designated 'environmentally sensitive areas' to be protected from the more extreme farming methods in order to preserve their flora, fauna and landscapes. Sooner or later such widespread disgust will lead to reform either through legislation or, in our market-dominated economy, by a major consumer switch to foods produced by more benign methods. The demand for organically grown food has been increasing for some years and I believe this trend will continue and accelerate.

An increasing number of farmers will be compelled to modify their methods. Most are unlikely to move so far as to qualify for an organic label, but change they will. It could be a process that moves in stages. The less intensive might seek to become fully organic, the more intensive to become less so for a time, and then to follow their colleagues. One way or another it will amount to a major shift toward organic farming and horticulture.

Farmers may find they cannot afford not to make the change, but can they afford to make it? Is the real choice not between conventional farming and organic farming but between conventional farming and bankruptcy?

It is a basic tenet of economics that any change is financially undesirable. This is the charge most often laid against those who encourage reforms of any kind. In this case, however, other income may be available to farmers to bridge the gap. Former town-dwellers who have migrated to villages may not like the idea, but the demand for recreation in the countryside is already leading many farmers to provide facilities that pay their way. Elsewhere, farmers might be employed to manage land in particular ways in the public interest. Rural communities must be allowed to survive and if farmers are to be asked to perform the necessary task of land management, rather than to grow food, it seems only fair to pay them.

The pressures that bring this change are also the factors that make it feasible. In the 1940s, even as late as the early 70s perhaps, the perceived need was to increase domestic production. Now, as I have pointed out, that need has faded. Our membership of the EEC provides security, the world at large is a much less threatening place, and within Europe we grow too much food. The idea that organic farming may be somewhat less productive in terms of crude output now holds no terrors and is far outweighed by the environmental benefits that would accrue.

So how should farmers begin the transition to regimes that depend much less on pesticides? They will need to learn more science so they can farm more scientifically. This answer will seem paradoxical only to those who equate organic farming with a kind of optimistic neglect.

The first step is to acquire a much deeper knowledge of the ecology of field crops. This involves detailed study of the populations, including the soil populations, that are found in the field and the relationships among them. It means investigating the biology and behaviour, and especially the reproductive and feeding strategies, of 'pest' species and also of species that act, or might be induced to act, as predators upon them. Notice that from this point the words 'pest' and 'weed' must be confined in quotation marks, for their meaning is only economic. Scientifically, 'pests' and 'weeds' are merely opportunist species able to exploit the abundant resources farming provides.

A consequence of such studies may be a more precise, more

limited use of pesticides. A species that conventionally is subjected to spraying repeatedly throughout the growing season of the crop might be controlled by spray applications only at those points in its life cycle when it is vulnerable. The pea moth (*Cydia nigricana*), for example, overwinters in the soil as a larva wrapped in a cocoon, pupates in late spring, emerges as an adult in summer, mates almost immediately, and lays eggs. The eggs hatch, the larvae search for young pods to penetrate and after three weeks they emerge, fall to the ground, spin their cocoons and become dormant. They are vulnerable only in the brief interval between the time the eggs hatch and the larvae enter a pod and again between the time they fall to the ground and disappear into the soil. As the peas begin to flower, the scientific farmer sets out traps with sticky floors, baited with the sexual attractant by which pea moths locate mates. When the traps start filling it means the moths will shortly be laying eggs. One spray application a week later will catch most of the larvae before they enter pods and a second, three weeks after that, will catch the survivors as they fall to the ground.

In other cases it may be necessary to preserve a population of a prey species that has a potential for 'population explosions' to provide a food source for the predators that control it. When it expands, so do they and problems are prevented from getting out of hand. Were the 'pest' eliminated completely the predator might leave to seek its fortune elsewhere. Many insect species overwinter on certain plants, so the removal and destruction during the winter of those plants from the vicinity of fields reduces the population dramatically.

Some techniques are well known to most farmers, but could be developed further. It is no secret that many insects prefer to spend the winter on certain plants but it might be profitable to find out how many more insects live in this way, and on which plants. Other strategies, like that with pea moths, work with some species but not others. It would be desirable to increase the list of those to which it might be applied. There may be other ways to favour predator populations.

Where pesticides continue to be used, even in the conventional way, it should be possible to achieve very substantial reductions in the quantities of chemicals involved. Ultra-low-volume sprayers, for example, were invented for use in Third World countries where the cost of agricultural chemicals has always been a limiting factor in their use. The hand-held sprayer uses torch batteries to power a small motor that spins a serrated disc at high speed. A concentrated solution of pesticide is fed to the centre of the spinning disc, flows centrifugally outward and along the serrations, and leaves the sprayer as a fine mist all of whose droplets are the same size. As a mist, it drifts into the crop, coating all sides of all plants evenly, but without dripping to the ground. It can achieve the same level of control as an ordinary knapsack sprayer, but with one-tenth to one-hundredth the amount of pesticide. It can also be used in the same way to apply foliar feeds.

It begins to sound as though I see a future in which, in addition to the now ubiquitous computer, every farm has a small but well equipped biological laboratory. I see nothing wrong with this idea, nor with the idea that biologists in general and ecologists in particular find employment either directly on farms themselves or with firms that provide a scientifically based advisory service to farmers.

We are the product of our past and the past is where any serious study of the future must begin. Remembering yesterday, then, let me speculate about tomorrow.

When I was a small boy the people of this country came close to starvation. Learning from the experience, it was decided that, so far as possible, Britain must never again be reliant on imports for its food. In those days, the prizes went to anyone who could devise any means of increasing output. As time went on and output rose, perceptions began to change. Problems appeared, some real, some imaginary, making more and more of us wonder whether this drive for output at any cost had perhaps gone far enough. Now, when our food supply is secured, we can afford to urge moderation in farming methods, and if necessary to demand it, in order to protect other things we hold valuable.

In order to carry farming forward, into what will amount to a new revolution in its methods, we must build on its scientific base. We no longer accept the crudity implied by drenching whole fields with costly and possibly harmful chemicals when more knowledge, more skill and more care could achieve as much or more by subtlety. It might not mean the total abandonment of pesticides, but it would surely lead to a very great reduction in the frequency with which they were used and the amounts involved.

The future, then, as I see it, leads to a more truly scientific farming. It will be gentler, less disruptive, and more efficient. If there is some loss in output in the short term it will not lead to shortages and so far as farmers are concerned there are ways to cushion the financial consequences to them. In the longer term I see no reason why farms should not be as productive as they are now, and profitable to the farmers.

We may not have unlimited time to make the transition, however. It may be another 10 years before we know certainly whether we have so modified the climate as to produce a general warming, and we may not know even then just what the effect may be for this country. It does not follow that a global warming means a warmer climate everywhere, and Britain could become cooler or could retain something much like its present climate. If it does grow warmer, though, we will find ourselves growing new kinds of crops and dealing with new species of opportunists. When and if this happens our farmers and their advisors will need all their skills. It is not a moment too soon to begin acquiring them.

Chapter Five

The Need for Integration—Farming and the Environment

Fiona Reynolds

AGRICULTURE AND THE FUTURE OF FARMING rarely seem to have been more vulnerable. Whether it is successive food quality scares, lamb wars with France or the international negotiations of agricultural subsidies under the General Agreement on Tariffs and Trade (GATT), no one can have missed the fact that the early 1990s have seen agricultural policy in crisis.

In fact, the pressures have been building for some time. Throughout the 1970s, public awareness of the damage wrought by intensive agriculture on the environment mounted steadily. The 1980s saw rising public and political concern as food surpluses mounted and the cost of the European Community's Common Agricultural Policy rocketed out of control, while the turn of the decade saw food scare after food scare affecting (or so it seemed) each sector of the farming community in turn.

Yet there has seemed little desire to tackle these problems in a coherent manner, and to set agriculture on a new footing, more appropriate to the needs of society as it moves into the 21st century and very different global circumstances. To take but one example, the Government's environment white paper, published in September 1990, ignored the long-running tensions between agriculture and the environment and treated the debate as one in which solutions had largely been found and problems were barely noticeable. To most commentators, including those within the farming industry and many farmers themselves, this was remote from reality.

Instead of establishing a decisive new direction for agricultural policy, the Government seemed to be content to let the industry lurch from crisis to crisis. To many farmers that means steadily declining farm incomes with little prospect of greater rewards except through greater capital investment or intensification. To the public it means an agricultural industry that continues to be unsympathetic to wildlife and the countryside and to be too heavily dependent on chemical and other artificial inputs with all their attendant concerns. Can things go on this way? Surely not. But to answer that question, we have first to address the concerns underlying the debate about the future of farming.

Agricultural policy has been one of the most consistent and (within its remit) successful post-war policies in the UK. Food shortages during the war and a commitment to produce more from our own resources led to a highly successful drive to increase agricultural productivity and efficiency. The cost per unit produced was driven downwards through the reduction of labour inputs, through the application of chemical and other artificial fertilizers and pesticides, and through rapid mechanization.

Generous financial incentives were offered to farmers to pursue this course, in the form of guaranteed prices for output and capital grants for investment. Under such single-minded direction, it is not surprising that dramatic improvements in efficiency were achieved and that in the process large swathes of countryside were transformed by the twentieth-century agricultural revolution.

The results of that transformation have been well charted, but bear repetition because they can still strike horror into the hearts of those who care about the future of the countryside:

Wildlife habitats lost since 1947:
50–60% of lowland acid heaths destroyed;
30% of upland heaths lost or damaged;
30–50% of ancient woodlands grubbed up;
50% of lowland fens drained and ploughed;
95% of lowland hay meadows destroyed;
80% of lowland permanent grassland lost or converted.

These figures did not emerge until 1983, when the Nature Conservancy Council published its first Great Britain Nature Conservation Strategy, charting for the first time post-war losses to important wildlife habitats. Shortly afterwards, the Countryside

Commission (which is responsible to the Government for landscape protection and public enjoyment) commissioned a survey of losses to landscape features, showing not only that a staggering 109,000 miles of hedgerow have been lost since 1947, but that losses had actually accelerated during the 1980s:

1969–80: 2,900 miles per annum lost;
1980–85: 4,000 miles per annum lost.

There have also been worrying changes within the social structure of agriculture. Since 1947 the number of farms has fallen sharply and there has been a steady shedding of labour even in recent years. For example, the number of full-time farmers in the UK fell by 10 per cent between 1978 and 1988. Production has tended to be concentrated on the larger units, with smaller farms struggling more year by year to stay in business.

But it did not need statistics to alarm the public, which, throughout the 1970s and 1980s, grew daily more aware of the conflict between an ever-increasing emphasis on efficiency and production in agriculture and the protection of our national inheritance of beautiful scenery and a rich wildlife resource. National *causes célèbres* such as the ploughing of moorland on Exmoor and the North York Moors and the drainage of the Somerset Levels and Halvergate Marshes, were illustrated poignantly by equally distressing local losses—a favourite hedgerow grubbed up, a pond filled in, a herb-rich meadow ploughed.

At around the same time as public concern was rising over the damaging impact of intensive agriculture on wildlife and the countryside, hackles began to be raised over the sheer cost of the Common Agricultural Policy. Food surpluses in a wide range of commodities became embarrassingly large, costing the Community a third of its entire agricultural budget in 1986, simply to store and dispose of them. Milk quotas were hurriedly introduced in 1984, but the problem was much more widespread, as the bulging grain stores throughout the middle years of the 1980s firmly indicated. The irony was sharpened by the fact that at the same time as Europe's food mountains reached their height, hundreds of thousands of children were dying of starvation in Ethiopia. This could not but add to the sense of urgency about the need to address the problem.

A flurry of activity in 1987 appeared to see Europe's agriculture ministers grasping the nettle. A new financial regime was introduced—

called stabilizers, because its intention was to stabilize the CAP budget—whereby an upper limit was set on Community grain production, with financial disincentives introduced to penalize over-producers. Early in 1988, supply control mechanisms were extended from the milk sector into grain, with the first European set-aside policy aiming to cut grain production by 20 per cent.

Yet, in practical terms, these reforms were more apparent than real. The stabilizer mechanism as a means of reducing production largely failed, since production targets were set too high and the financial penalties too low to discourage producers. The only tangible effect of these first tentative steps towards budgetary control has not been a reduction in the cost of the CAP itself, but a steady decrease in commodity prices at the farm gate, leading to falling farm incomes of at least 25 per cent in real terms.

The much-vaunted set-aside policy has had the effect of reducing production throughout the Community by the grand total of less than 1 per cent. This is due to a combination of low take-up and inadequate anticipation of the 'slippage' phenomenon—where farmers who set aside some of their land increase production on the rest to compensate. Implementation has been less than enthusiastic in many member states and the policy less than popular, not least because the scheme was introduced in a great hurry and failed to provide even basic environmental and good husbandry safeguards. As a result, set-aside has attracted opprobrium not only from members of the public (who see the scheme, rightly, as paying farmers not to farm) but from fellow farmers, too, as untidy, weed-infested fields have caused problems for neighbours.

Indeed, the only reason grain surpluses in the Community have apparently subsided is because a series of poor harvests elsewhere in the world have allowed EC grain to be disposed of cheaply on international markets rather than languishing in expensive grain stores. We still produce too much and, as soon as world harvests pick up, will be faced once again with embarrassing food mountains.

Following its feeble attempts at internal reform, the Community has had to face what may turn out to be more important long-term international challenges in the form of negotiations on the General Agreement on Tariffs and Trade (GATT). The highly protectionist nature of the CAP has for a long time enraged other nations, especially the United States and the Cairns Group of trading partners, all of which have instituted significant reforms to their own agricultural policies.

Their demands for swingeing cuts in the Community's guaranteed price system precipitated a crisis in the GATT process, which in turn served to highlight what everyone had long known: that it is no longer ideologically, economically or environmentally sustainable to run an agricultural support system which is premised on paying farmers a guaranteed price for food whatever they produce and however they produce it. If the GATT process did nothing else, it demonstrated the need for a more coherent and responsible rationale for supporting agriculture.

That rationale needs to rest on a more broadly-based set of objectives for agricultural policy. Within Europe, at least, we can no longer endure the cost or political embarrassment of food surpluses—so the overriding emphasis on production must end. Instead of the single-minded pursuit of efficiency and higher production, we must integrate environmental and social goals into the mainstream of conventional agricultural policy objectives. This will require new mechanisms—to encourage environmental good practice and to regulate bad and to stimulate land management practices that keep people on the land, doing rewarding jobs which fulfil combined agricultural, environmental and social objectives.

Designing an agricultural support system to achieve these new objectives is, though, far from easy. First, there is the problem of achieving political consensus within the Community, whose farming industry varies enormously between north and south. The British Government has frequently been in the vanguard of economic reform, mainly to achieve cost savings, and is highly critical of the lengths to which other rich northern Governments, such as France and Germany, are prepared to go to protect their farmers. And, as so often, the weaker voices of the southern member states, whose farming industry is less resilient and efficient (though in general far more environmentally benign), must be content with simply accepting a slower pace of reform than that adopted by the stronger member countries.

Any new agricultural system must reflect these realities as well as being sensitive to the enormous variety of landscapes, farming types, cultures and socio-economic conditions in the Community.

But, assuming political consensus can be achieved, what should be the principles of a new agricultural policy for Europe? There are, put simply, four central requirements.

First, the *price support system* must be cut so that prices for agricultural commodities are at levels approaching, if not at, those of world markets. High guaranteed prices have been the primary cause of much environmental destruction. When higher output is directly rewarded by higher income, there is a very clear message to the farmer to produce more per acre. This applies even when supply control measures, like set-aside, are introduced. Unless the whole farm is taken out of production, the availability of high guaranteed prices for produce from non set-aside land will encourage intensification on that land—hence the problem of slippage.

Intensification of production is the single biggest contributor to environmental damage. Landscape features and wildlife habitats are lost as every possible part of every acre is brought into production, more chemical fertilizers and pesticides are applied, and heavier and more destructive machinery is used. Thus, price cuts are not only increasingly inevitable as international pressures on the EC agricultural support system grow; they are a desirable and necessary vehicle for achieving an environmentally responsible agricultural policy.

However, price cuts on their own will not secure environmental or social goals. The effects of farm-income squeezes are already being felt in many parts of the community and can cause environmental problems of their own. In extreme cases, bankruptcy of an individual farmer might lead to the land being abandoned altogether. By far the more likely result in many parts of the community, however, would be that other land users would move in. Key contenders would include commercial forestry, building development, or other, more efficient farmers whose tendency would be to farm the land more intensively than the previous occupant. All of these outcomes would be more damaging than maintaining the existing farmer on the land, from both an environmental and social perspective.

Thus, the second principle must be that *price cuts must be accompanied by a new form of direct payments to farmers in return for positive environmental management.* In other words, recognizing that the market is inherently unable to place a value on or reward the environmental benefits which sensitive farming produces, society should agree to pay for them directly. A system of this kind, while capable of widespread application, would be especially valuable in those parts of the Community and within individual member states where agriculture's destructive influence has been less profound and where payments for the retention and management of features and areas of environmental quality, which would otherwise be under threat from intensification, are particularly important.

A system of Environmental Management Payments (EMPs for convenience) would need to operate through the guidance budget of the CAP and could be funded by the savings achieved through reductions in price support. The principle of EMPs would be to provide the farmer with an annual payment based on the environmental assets of the farm, in return for the farmer's agreement to manage these in ways which maintained and enhanced their environmental value. What would be revolutionary about this approach is that it would not be confined to special areas of countryside, but would be available to any farmer, anywhere, provided he agreed to undertake the management obligations.

Of course, it would be open to an individual farmer to operate his farm wholly on market principles, being rewarded only by the new, lower prices available for what he produced. Some highly efficient farmers might wish to do this, but the majority would find the effect of lower prices without any compensating source of income untenable. It would seem probable, therefore, that EMPs, properly constructed, could appeal to a very wide range of farmers and farm types and bring substantial benefits to the environmental quality of the farmed landscape of the Community as a whole.

The introduction of a scheme of this kind—simple to administer, clear in its objectives and capable of very wide application—is essential if we are to move away from the treatment of environmental issues as marginal green fringes. To reinforce this point, an EMPs scheme would need to be complemented by effective environmental controls over, for example, hedgerow removal, pesticide use and other potentially damaging activities. A special-areas approach, confining environmental responsibility to, say, Environmentally Sensitive Areas or Nitrate Sensitive Areas, actually makes the prospect of more fundamental reform farther off. Only by making the issue of environmental responsibility one of central relevance to *every* farmer, whether in the arable prairies of East Anglia or the remote hills of Northumberland, will the environment take its proper place in agricultural policy.

Having established the central importance of a fairer balance between the market's ability to reward a farmer for food production and society's responsibility to reward him for his countryside management skills, the third priority is to consider the *quality of food production* from the land.

Food scares in recent years have done much to bring public attention to the issue of food quality and have resulted in a much better informed and more demanding consumer. This can only be a welcome development and should put pressure on producers to justify the level (and in some cases the principle) of artificial inputs to crops. The question of food quality raises, in particular, the prospects for organic farming, whose products command a premium because of their guaranteed freedom from artificial fertilizers. The market share of organic produce is increasing and should do so further as consumer awareness grows.

However, because organic produce is associated with the idea of environmentally responsible farming, it is vital that the standards which govern the industry reflect a wider definition of environmental concern than simply the application of chemical fertilizers or pesticides. It is of no net environmental benefit to produce a field full of organic carrots if the field used to be a flower-rich hay meadow and its hedgerows have been replaced by post-and-rail fencing. The Soil Association's symbol achieves this objective, by embracing landscape and wildlife protection standards along with those of animal husbandry, as well as having strict rules about inputs to the land. UKROFS (the new UK Register for Organic Foodstuffs) has yet to take this step, which involves recognizing the vital importance of securing environmental credibility across the board in the new standards for organic food.

The opportunities and market share for organic farming are growing, and the Soil Association's objective that 20 per cent of Britain's land should be organically farmed by the year 2000 looks increasingly feasible. However, even if this—or a more ambitious target—is achieved, it is both unlikely and impractical that more than a significant minority of British farms will be farmed organically in the early years of the next century.

There is, therefore, a need to recognize that an important bridge between organic and 'conventional' farming deserves greater consideration. Thus, the fourth requirement, and one which complements and reinforces the first three, is to develop what are becoming known as *'integrated' farming systems*, which aim to operate on environmentally sustainable lines, but which do not go as far as achieving organic status. Integrated farming is, like conventional farming, highly technical and specialized, but takes a long-term view of the future sustainability of the agricultural industry and aims to minimize artificial inputs to the land and energy-intensive or artificial products in favour of more environmentally friendly ones. For example, experiments in the Netherlands have shown that pesticide applications can be reduced by over 90 per cent without a loss in farm income, by substituting animal manure for artificial fertilizer, and by using mechanized weed controls and less vulnerable plant strains.

There are existing mechanisms in place which could be used to promote types of integrated farming that could be environmentally beneficial. The EC, at the same time as it introduced the set-aside policy, also made a commitment in principle to 'extensification', whereby farmers who achieved a 20 per cent reduction in output across a whole farm would receive annual payments. The extensification regulation (currently confined to pilot projects, because of the claimed difficulties of administration and monitoring) is particularly promising for those parts of the Community where intensification has been rapid and widespread and where considerable benefits could be achieved from a reduction in yields of the order of 20 per cent.

The implementation of these four principles may sound ambitious or even plain star-gazing. But the truth is that we are nearer to a breakthrough in reorienting agricultural policy than at any time in the past. International pressures have exposed the unsustainability of the old EC system, public concerns about the environment, food quality and the role of farming in protecting the countryside are at their height, and farmers themselves are only too well aware of the vulnerability of their position and the need for a new agricultural support system which can command greater public favour.

Sadly, the record of reform is not encouraging. Time and time again, the Community has claimed success in economic reforms, only to have its efforts exposed a year or so later. It claims to have been 'greened', only to be shown that policies for special areas are too weak and leave the vast bulk of expenditure still directed to environmentally damaging activities. It claims to support farmers and farm incomes, without acknowledging the steady stream of farmers from the land and into bankruptcy throughout the CAP's history.

But, in environmental terms, the risk of *not* reforming the CAP is too serious to contemplate. Farming is undoubtedly the most important and influential land use in the Community. Its activities affect between 70 and 80 per cent of the surface area and, while it has caused untold damage, it is also the central means whereby the future protection of the bulk of the Community's wildlife and landscape resource can be secured. It is time to get farming policy right and we cannot afford to wait much longer.

These ideas could provide a framework that combines political realism with practical application, but which does not compromise on the fundamental nature of the shift in orientation of the CAP that needs to take place. The next five, 10, even 25 years will be crucial in determining whether decision-makers in the Community are ready to grasp the nettle of agricultural reform boldly, or are prepared to live with the current uncertain, incoherent and discredited system.

(Further details of the EMPs proposal are provided in the CPRE/WWF publication *Future Harvests* (November 1990) available from CPRE, price £6.)

Chapter Six

A Policy for Forestry and Woodland

Penny Evans

IN ANCIENT TIMES Britain was as well wooded as the rest of northern Europe and early farming communities of the Iron Age would have spent considerable periods clearing the trees and shrubs. This work was essential not only for agriculture but to provide timber, which formed the basic resource for tools, shelter and fuel. And so it continued, with woodland being tamed to provide the required types of timber; different species, sizes and shapes depending on future use, or being cleared to establish open land for farming and for settlements. Nowadays, only small remnants of the original woodland remain, altered by human management but still recognizable through the particular complement of plant species. The vast majority of woodland present today reflects, however, much more recent forestry fashions: the nineteenth-century plantings of exotic North American tree species; the large estate woodlands with good stands of oak and beech; and the more recent intrusion of very large conifer blocks found principally in the uplands, a direct result of Government forestry policy since the First World War.

Today, Britain cannot be regarded as well wooded. With only 10 per cent of our land under trees we, along with Ireland, have the least woodland of any European country. In England the figure is worse, with less than 8 per cent woodland, of which about 22 per cent is ancient and semi-natural—the likely wildwood remnants. Included in the total there is also hedgerow or small blocks of relatively unproductive woodland, as well as modern plantation stretching mile upon mile through the hill country. Many of us, whether forester, environmentalist or sentimentalist, would like to see more woodland. Yet, over recent decades, foresters and environmentalists have been in conflict, not because of any disagreement about wanting more trees, but rather over the way in which forestry has been carried out.

Traditional forestry, practised over the last centuries by one generation of woodmen following another, reflected the variety of soil, climate and landscape as well as the wide range of uses for the wood and timber produced. Coppicing of hornbeam, hazel, ash and lime provided small timber for fuel, for tools, for charcoal; oak, ash and beech grew to maturity for building houses, barns and boats, while tannin was produced by coppicing oak. Pollarding of trees, at shoulder height, provided wood and fodder in parkland as well as protection against grazing by cattle or deer. The scale of operations was small and new trees were established through natural seeding, careful selection and weeding. In such a way the woodman bent his forest to provide the crops he required for the wide variety of uses for which timber was then the only resource. Nowadays plastics, petrochemicals, alternative fuels, new crops and mechanically produced timber products have reduced the need for structural and species diversity.

But it is forestry policy set out by successive Governments since the early twentieth century that has done more than anything to create conflict between foresters and conservationists. Following the First World War, the Forestry Commission was established in order to stimulate afforestation and reduce the reliance on imported timber, which had developed with the expansion of the British Empire. Lacking timber during the war, the Government realized positive action was needed, following the large-scale fellings found necessary during the war, to encourage further planting. So the Forestry Commission commenced work with a policy intended to increase our timber stocks in the fastest possible time.

During the first few decades of the Commission's existence, planting, principally with a variety of conifer species, took place on land purchased by the Commission and, through the provision of grants to private landowners, on estates throughout the country. Only following the Second World War was attention focused primarily on hill land. This was the result of widespread concern over agriculture which, in decline before the war, only just managed to provide enough food for the duration of the war. In the following years agriculture was expanding in an attempt to ensure that Britain was as self-reliant as possible in foodstuffs. As a result, forestry was pushed to land of low productivity.

It was very difficult to promote forestry when only the poorest land was available for planting and the remaining lowland woodland was being cleared to provide a greater area of agricultural land. Small woods, hedgerows and individual trees were removed to allow increased agricultural efficiency. As a result, forestry policy was developed to provide substantial grants and tax advantages to stimulate afforestation. These incentives were particularly appealing to wealthy estate owners with a heavy tax burden. Forestry was justified not only in import savings but by its employment potential in remote rural areas. A shaky economic justification was promoted, suggesting that the faster growing, exotic conifer species—in fact, almost the only species with any chance of growing in the uplands—provided a real return on investment. These economics took little account of the ravages of storms or pests, or of soil and water depletion, and failed to consider the cost to our landscape and wildlife.

Forestry had been forced away from its earlier integration with agriculture to the only parts of Britain where agriculture was declining in productivity. The result was that the type of forestry became very restricted, because of the physical limitations of the site as well as the reduction in the variety of timber required.

With hindsight, the Government and the Forestry Commission may well have been justified in their action immediately post-war but they became locked into large-scale monoculture of Sitka spruce, lodgepole pine or larch, fuelled principally by tax exemption. As the power of vested interests in this tax exemption grew, so the opportunity for changing the direction of forestry diminished. At the same time, concern was growing over the damage occurring to our remaining undeveloped landscapes and unspoilt wildlife habitat with the increasing pace of change in the countryside. An inevitable clash resulted as afforestation was creeping over the wild areas of Scotland, England and Wales, damaging scenery, habitat and natural beauty. This clash has still to be resolved, for large-scale conifer planting on hill land continues, although now almost solely in Scotland.

Today, we have inappropriate planting justified by incorrect assumptions and supported by misdirected policies. We have lost the natural diversity as well as the man-made variety found previously and we continue to lose important semi-natural areas. In return, we have a system of forestry which is dependent on deep ploughing of steep hillsides, application of fertilizers and pesticides and the planting of non-native trees grown in tree nurseries from limited genetic stock. All of this is supported through dubious economics and limited social benefits and promoted by policies which set open-ended targets not of productivity, quantity or quality, but of the raw materials used, the area of land planted. What other industry measures its success in quantity of raw materials consumed rather than output?

The final irony is that most of us believe trees to be a good thing. We support the Woodland Trust and the Tree Council; Men of the Trees sounds laudable and the job of forester seems sound and healthy. Yet the implication of continuing our programme of forestry and afforestation without radical change to both practice and policy is further discontent between foresters and those concerned with protection of the environment. The latest dubious justification for afforestation is that trees mop up carbon dioxide, the principal greenhouse gas. This is nonsense, because the deep ploughing and draining of the hill peat land releases masses of carbon dioxide as the peat is oxidized, so the initial balance is in fact heavily towards increases in carbon dioxide.

There is an alternative. Forestry could enter the next century with renewed vigour if the fresh opportunities arising from agricultural change, wider environmental concerns, and widespread public interest and support are seized and if forestry policy is radically altered to provide the framework and financial support for the new direction.

The most positive contribution forestry can now make is in the lowlands that have been scarred by intensive agriculture but are now open to a wider range of land use and methods of management. With entry into Europe, technological change, and sustained damage to our soil, water and atmosphere all directly affecting agriculture, there are unprecedented opportunities for a forestry which is not based on large blocks of monoculture. It should reflect the broad spectrum of purposes for new woodland—improvement of landscape and wildlife habitat, protection of water catchments and soil, opportunities for recreation and public appreciation—as well as encourage renewed management and the expansion of our existing woodlands.

What then of the future? It is certain that we can no longer allow damaging afforestation and forestry practice to continue the onslaught on our remaining semi-natural land. In England, this was recognized explicitly in 1988, when the Government announced that no further large-scale afforestation of the English hills would be permitted, specifically in order to protect the remaining unimproved upland habitat. This local victory came 50 years after the initial skirmish, when the Forestry Commission and the Council for the Preservation of Rural England reached their historic 'agreement' to protect the core of the Lake District from conifer afforestation. Will it take another half century to halt afforestation of the remaining Scottish hills? Even today, extensive plantations of conifers are being encouraged with financial incentives and with minimal consideration of the environmental impact or consequences. A recent application for 873 hectares has been approved, of which 807 hectares are to be planted with spruce and larch and the remaining 66 hectares (just over 7 per cent) with native species. This scheme is not alone and such programmes continue despite heavy criticism not just from the environmental lobby but from Parliament[1], from economists[2], and foresters[3]. The criticism is based on a range of issues, not just environmental damage. The economic justifications for this type of conifer afforestation were discredited in the National Audit Office Report of 1986 and the social benefits of maintaining forestry in depopulated rural areas have rarely been of significance, but so far, the power of vested interests—principally the landowners and their associated forestry companies—has won the day.

Nevertheless, it is not possible to be entirely gloomy. New opportunities abound and both Government and individual landowners and managers are seizing them. There is the widespread support for increasing the area of woodland in the lowlands in order to restore damaged landscapes, to provide an alternative productive land use to agriculture, to explore potential recreational activities such as horse riding, walking, and orienteering, and to increase the wildlife habitat. Grants are available to farmers for all these activities, for the protection of the specially important ancient woodlands, and soon for the management of existing woodland. However, it is the cost of land, inflated over recent decades by agricultural subsidies and financial arrangements, which still provides the biggest obstacle to changing our pattern of afforestation.

There are also the much tighter restrictions on the management of existing woodland, including more sensitive conditions for woodland operations agreed through the current consent-and-grant procedures operated by the Forestry Commission and local authorities. Tree preservation orders are likely to be supplemented by hedgerow preservation orders, felling licences are not approved for clearance of woodland, and woods are expected to be felled in sections, rather than by the blanket clear-fell of the past. Planting grants encourage the use of broadleaf species and ancient woodland is now more carefully identified and specially managed. In these arrangements, the official and voluntary countryside conservation agencies play an influential part, for without them the environmental dimension is often, in practice, still secondary to timber production goals.

Finally, there are the new initiatives which themselves reflect the potentially widespread support for a fresh approach to forestry; in particular the community woodlands and forests, which are intended primarily to meet local community objectives, re-establish countryside on the edge of urban areas for peaceful enjoyment, enhance the landscape and disguise old scars, and encourage interest and participation in tree planting, natural history and woodland management. The largest of these schemes is the proposed New Midlands Forest, an innovative and imaginative concept seeking to meet social, environmental and economic objectives, but which has yet to be put into practice. These projects will require both the political will and, more especially, a financial commitment from Government, for again the distorted agricultural market, reflected especially in land prices, causes severe difficulties. There are the successful range of small woodlands projects which, backed by local authorities and countryside bodies, co-ordinate and stimulate the restoration, management protection and careful exploitation of existing small woods. These projects seek to provide a vital link between woodland owners and the practices of management and marketing.

Into the 1990s and the next century, there is the urgent need to re-evaluate our actions, to take account of the effects we are having on our global, national, regional, local *and* personal environment. In many spheres, this means cutting back on consumption and exploitation and properly assessing the impact and costs to the environment. With our woodlands, however, we have the potential for tremendous benefits brought about by this reassessment. We can look to wider use of timber products from sustainable resources. We can reduce our reliance on both imported timber and on man-made substitutes. We can look for the benefits of woodland to water catchments, soil stabilization, locking up carbon, and to the ecology and protection of our countryside. All of these give the real environmental return of appropriate forestry, which is not gained from the current focus on species monoculture, ploughing of peatlands, and reliance on planting rather than on natural regeneration. An ambitious programme of afforestation on intensively farmed land alone could, however, lead to the criticisms levelled at the current afforestation programme in Scotland.

The essential element of our new programme has to be flexibility: flexibility in restoring the variety of woodland in hedgerows, field corners, and small woodlands; and in supplementing existing woodland and larger forests. In each case regional woodland characteristics should be respected, including the range of different species and different management techniques which avoid uniform practices and focus on enhancing our landscape.

This future for forestry can be brought about only by clear direction from the policy makers, for forestry has always been heavily and directly subsidized from the public purse, by support from the general public and by a commitment from the forestry lobby and industry. We must first recognize the problems associated with current British forestry policy and practice, identify (and to a degree quantify) the sources of benefits that can be tapped, and, finally, prepare a forestry policy, programme and package of projects for implementation.

In practice, this requires three key changes. First there should be a wider commitment from Government to the real benefits of forestry in Great Britain in the form of a clear statement of new objectives. Secondly, forestry support and supervision should be moved away from a single separate agency, the Forestry Commission, perhaps to an integrated countryside management authority. Then the wider costs and benefits of forestry can be incorporated with agriculture to provide for a balanced land-use strategy which addresses landscape protection, nature conservation and recreation as well as food and timber production. Thirdly, planning authorities and parish councils, voluntary bodies and local interest groups, should be able to contribute to decisions regarding the fate of our countryside. We must also insist on silvicultural practices which favour selective felling of woodland rather than extensive clear-fell, concentrate on natural regeneration to restock woodland, give attention to non-timber species, such as the underwood, shrubs and ground flora, ensure care over the timing of woodland operations, bring about a marked decline in the use of chemicals within woodland management, and encourage a return to community involvement in tree and woodland establishment and planting.

In Britain, we have supplemented our timber requirements through the over-exploitation of the tropical forests with a resulting deterioration in our own forests and damage to the natural resources and the economic well-being of the developing countries. It has been our demands that have led to a number of the developing countries selling too-valuable tropical timber, while ignoring completely the longer term damage to soil fertility, to species diversity, to the culture of local communities, and to the global climatic balance. We in Britain can no longer avoid recognizing our contribution to these impacts. By showing that we recognize and practise sensitive, sustainable forestry we can help persuade others to do likewise.

Chapter Seven

The Sustainable Successes of Permaculture

Graham Bell

PERMACULTURE IS A DESIGN SYSTEM for creating abundant human habitats. It tells us how to place all the elements in our living space so that they interact to create the maximum beneficial yield. It is derived from the words *permanent* and *agriculture*, but also implies permanent culture. Why? What is so important about permanence? And, if it is important, how does Permaculture itself work, and can it succeed here, now?

Anything that isn't permanent is to some degree temporary, so what do we accept as being a long time? The life of this parliament, or a geological era? We talk arrogantly of things as 'dinosaurs' which have outlived their usefulness. Dinosaurs, of course, occupied the Earth for many millions of years longer than *Homo sapiens* has done. We do have this habit, we people, of using language to blind ourselves. *Sapiens* means wise. What's the evidence for naming ourselves so appreciatively?

If you do not already ask questions like these you will probably not accept that Permaculture is needed as a solution, because you will not accept that the problem exists. We live in a world in which fewer people are getting richer and more people are getting poorer. Population is escalating and the biosphere, the living outer skin of our planet, is groaning under the strain. We are pumping out pollutants at a rate greater than the capacity of the planet to absorb them and we are gobbling up resources faster than the world system can create them. You do not have to be a mathematical genius to calculate that we cannot carry on like this indefinitely.

Sustainable systems must create all their own needs and use all their own outputs. In other words, pollution is only unused resources and work is the self-creating task of people in systems designed without naturally regenerating inputs. If you don't have enough fuel you have to work to make it, or work to make money to buy it. Everything you produce and throw away is wasted energy. So, in a sense, Permaculture offers an infinitely adaptable system of energy management. We try to design life so that each source of energy is made to work within the system for as long as possible and in as many ways as possible.

Water is a good example to start with. A certain amount of water comes to each piece of land, as rain, snow, mist or through-flowing water courses. We can apply energy to drain land and speed the flow of water through the system or we can apply Permaculture thinking. Let's use the water. Let's use its characteristics as an asset; dam it as high as possible in the landscape as a reserve; make it drive turbines and generators; build houses by it to benefit from heat gain from reflected winter sunlight; stock it with fish and fowl; recycle it through biological cleansing systems (reed beds etc.); build a deep topsoil that will store it against drought; and make it drive ram pumps to return a proportion of itself to the top of the system.

Everything is an asset and should be viewed as such. All energy is precious. So it's important to have a scale of values. First, do those things that create energy (plant trees, for instance). Next, do those things that conserve energy (home insulation, dams) and, only lastly, use energy.

Permaculture offers a design system for making the choices needed to cope with a sudden, and possibly economically painful, change of direction. Every change implies a human intervention. Yet nature shows us this is often counter-productive, so our first option is *to do nothing*. If you cannot maintain yield by this means, or if predators get in, try *accepting some loss as natural* (cost it in!). If intervention becomes necessary, the best solution is a *biological* one, such as companion planting and use of green manures. If that doesn't work, try mechanical intervention (e.g. ploughing, fencing). Only as a last resort should we contemplate *chemical* intervention, whose consequences are so often unknowable. Permaculture as a discipline doesn't discount any techniques, but asks you to audit the energy cost of the technique before implementation. Bulldozers and JCBs may be the most energy-efficient ways to lay in irrigation channels and roadways, but 4.5 litre tractors are not the best way to build fertile topsoil.

If you look at history, you will see many great civilizations which flourished and fell. They all rose up on productive agriculture and forestry, and they all collapsed as their agricultures collapsed. Syria is a sparsely populated desert land which once was rich and productive. Greece waned in importance as it lost its topsoils; the Roman empire in North Africa faded as the earth became exhausted and the desert encroached. In Scotland, where I live, they have this year used snow-ploughs to clear topsoil from the roads, blown there by the wind. Anyone who has stood in East Anglia and seen sugar beet stretching treeless to the horizon can sense the ominous future we face if we do not mend our ways. Anyone who has come face to face with the spiralling figures for farm debt must know that agriculture cannot continue to meet the escalating burden of interest which travels hand in hand with increased mechanical and chemical inputs. The figures show that Britain's farmers have done a magnificent job since the war. They have done all that was asked of them—in 40 years they increased food production 100 per cent with a reduced labour force and a diminishing quality of agricultural land. The figures also show that energy inputs to achieve that production have increased 1,600 per cent. In other words, farming is eight times *less* efficient than it was in 1945.

This is the crux of the matter. We talk about yield, but we are not measuring true yield. Our national accounting systems do not take note of the true costs of inputs and outputs. The financial burden of cleansing East Anglian water supplies of carcinogenic nitrate levels was not paid when the bag of fertilizer was bought. The irreplaceable fossil fuels which created the short-term profit of the fertilizer did not include a levy to build renewable energy resources to replace it. Yield equals output minus inputs. This is not as simple as tonnages past the farm gate. The latter figure does not account for these other costs, nor does it examine dry-matter weights, nutritional values, poisonous residues, or, and this is most important, whether the crop has diminished or increased the capital value of the soils.

Nature itself has the capacity to create deep and enduring topsoils as the fundamental fertility of the planet. Organic growing techniques are vital to the preservation and regeneration of living soil, and the flora and fauna which serve the soil. Darwin showed that earthworm populations alone can create 10 tons of topsoil per acre, per year. Their contribution is significant. Bracken is not a curse, but a pioneer plant whose function in the ecology is to mulch depleted soils and to rebuild the phosphorus levels. Its natural successor is gorse which fixes nitrogen. Left to its own devices, the regenerating moorland, freed of the ravages of sheep, will set rowan and birch among the gorse to restart the forest which will climax in oak and pine. Observations like these, based on natural systems and a proper understanding of their function and purpose, can enable the farmer to make huge leaps forward in true yield whilst also returning our countryside to a heritage rich in wildlife and aesthetic beauty. By gentle techniques of soil enrichment our four million acres of acid upland could be returned to 50 per cent forest and still support their present population of sheep on the balance. Farmers gain as they measure the capital of their land, not just carcass weights which in any case are maintained.

Bruce Marshall, on 1000 acres in the Pentland Hills, has used a policy of liming followed by earthworm introduction, for a dozen years. Clover regeneration and the return of native woodland have increased his game levels, whilst his stocking rates have doubled and his stock health has greatly improved. More yield for less effort—a classic Permaculture pattern, all based on patient observation, a resistance to official disapproval, and a willingness to try, fail, try again and improve. We do not need to teach our young farmers how to mend tractors; we need to teach them how to see for themselves the workings of nature.

Arthur Hollins farms in Shropshire. He has been going against the flow for 50 years and more. He has at various times led the way in added-value dairy products, and used pigs as pannage-fertilizers, feeding on woodland nuts and returning the manure to the fields. He stands out now on the success of his foggage system. Stock are overwintered outdoors on a rich, deep, herbal ley that is never ploughed. They are allowed to browse at will, being resistant even to the poisonous charms of laburnum, which they nibble. Arthur's farm is profitable and he doesn't apparently raise much of a sweat working it.

The use of trees and hedges as animal fodder is very underrated here, yet each species may have something different to offer. Alder and gorse are rich in protein and as part of a balanced diet are particularly beneficial to young stock. Hazel aids lactation in feeding mothers and willow has been known for its medicinal properties for generations. Gorse and alder are also nitrogen fixers, feeding the soil. All of them act as windbreaks, moderating climate. In winter they protect animals from extremes of cold and aid weight-gain, while in summer they protect them from extremes of heat. Excesses of temperature change are thus avoided. All increase the biomass of the planet and reduce atmospheric CO_2, thereby contributing to the reversal of the greenhouse effect. All increase the amenity value of the countryside. There are the added benefits of hedging as boundary fences, fuel, wood crops and wildlife havens.

It is an important Permaculture principle that it is not the number of elements in a system which make it abundant, but the number of beneficial relationships between them. However, another key point in Permaculture is diversity. Monoculture offers advantages to the administrator, but it is anathema in nature. If only one element serves a certain function and it fails, all is lost. If the function is replicated by a diversity of elements, we can still carry on. Losing elms to Dutch elm disease isn't such a blow if you have oaks, limes, and beeches as well. Mixed farming is a much better hedge against pestilence and bankruptcy. A mixed countryside is a healthier one from all points of view.

This need for a variety of resources also fosters a much healthier wildlife (often an important economic crop as well) and has two other consequences of note. Our climate seems to be getting more extreme, and this is happening globally. It is increasingly important to have reserves which will see us through drought or flood. Secondly, by fostering diversity we preserve the wealth of our genetic reserves. A world in which six species provide 70 per cent of our food crops is a perilous one. In Britain, six varieties of one species of potato is a very slim bastion between successful harvest and crop failure or even, one day, hunger. There is no tablet of stone which says Britain shall never suffer the fate of other civilizations which waste their topsoil for the benefit of profit now. No western agriculture survives without state subsidy, therefore none is *actually* profitable.

Britain has only about 3 per cent remnant woodland. What morality allows us to protest at the felling of the Amazon rainforest when we do not cherish the forest here? But it need not be like this.

Bernard and Emma Planterose have, in two years, created a tree nursery at Duartbeg in the very north of Scotland which can produce 100,000 trees a year—and this on former marshland, within sight of the Atlantic Ocean. Many others are recognizing the need to regenerate our tree culture.

The integration of aquacultures with farming is showing considerable increase as a source of high-value protein output—be it crayfish, trout or carp. The return of free-range pig production also reinforces our understanding that intensive stock production in buildings is not only cruel but leads to high levels of ill health and mortality and an end product of inferior quality. Consumer pressure to have food which is grown in morally acceptable conditions and which meets their health needs is increasing the general trend to organics.

But the organic approach does not simply mean a cessation of chemical inputs. It means a sympathetic understanding of all these cycles and patterns. Slurry ceases to be a problem and becomes a valuable manure. It might also yield methane as a power source and heat from its storage tanks. An 'organic' tomato which has just jetted halfway round the world has no proper right to the title, given the profligate use of energy getting it to the consumer. So 'organic' also implies grown for local consumption, and that means appropriate financial systems where communities invest in their own needs.

This could be of tremendous benefit to the British farmer. Local marketing means close links with local buying power. In Japan, there are hugely successful urban/rural food clubs where the farmer contracts to provide to specification the household's needs. In return, the urban centre provides labour at peak periods. All gain. And it's all possible here. Our 60-odd city farms are part of the way forward.

The success of organic growing in Britain is dependent on a growth in popular understanding of the issues, among both farmers and consumers. The greater the marginalization of all-powerful retailing combines the better—for too long the farmer and the consumer have been marginalized by *them*. Education is vital.

Permaculture offers a mighty tool to understand the links which empower ordinary people to control their own lives. Good food is essential to a healthy body. A healthy landscape is essential to healthy world-life and human souls. We cannot afford not to be organic. The rate of change may seem small now, but the growth rate is geometric.

Permaculture offers a workable structure to people wanting to free this planet. It asks you, first of all, to accept responsibility for your own needs. Knowing what's good for everyone else isn't the solution; personal involvement is the best start. Consequently, Permaculturists agree to garden. Gardening teaches you observation of natural cycles of growth and decay and also gets you growing your own food. It's frightening how few farmers are also gardeners. Human-scale food production is not only healthy and fun, but a great learning process. Field-scale techniques of no-till farming become more possible when you've practised on a raised bed of winter greens. Beneficial plant associations for fertility and pest deterrence can be practised on a small scale to apply broad-scale. Soil becomes like an intimate friend. Weeds have new meaning as you start to understand their purpose in the scheme of things.

A neighbouring farmer complained that a field was 'bad land'. Its fine crop of docks tell me it's badly compacted. As I watched him drive his tractor across it again, to inspect a fence that could have been reached by a 30-second walk, I wondered at how divorced he was from his land not to understand these messages. Three months later rainwater still filled the wheel ruts he left. I feel Permaculture is simply common sense which is not common enough. Soon this ancient wisdom will be shared amongst us once more, and the 'bad land' will disappear.

On a recent Permaculture course, Bernard and Jonathan made an urban garden in our cobble courtyard in an afternoon. It could keep two people in herbs and salads for much of a year. It would fit in millions of back yards throughout Britain. We have talked here of a few Permaculture successes. Permaculture gardens are frequent in this country—I have taken many photos to prove it. Soon Permaculture farms and forests will be as common. Many of these projects are started by people who have never heard the phrase; they are simply using their own native wisdom to place things strategically in their system so that work is minimized, pollution abolished, and yield is abundant.

There is no more powerful mark to leave on the world than to have made a garden. I have every hope that the sorry lessons of our environmentally damaged present will guide us back to the path, the path to the garden of Eden. I firmly believe that Permaculture is an easily assimilable solution of global application that can take us there. Start today if you have not already done so.

Chapter Eight

The Politics of Organic Farming

Ron Davies

MY CONTRIBUTION TO THIS VOLUME is intended to give practical suggestions on how the political establishment can assist the growth in organic farming. It is not a manifesto, in the sense that I steer very clear of party politics. But it is prescriptive. I believe there are obstacles—one in particular—to continued rapid growth in organic farming. I am sure they can be relatively simply overcome. Politics, in the classic definition, is the art of the possible: an expansion of organic farming is certainly possible and, according to many surveys of public opinion, would be popular.

The commissioning of this book reflects the fact that agricultural policy is in ferment and, for the first time in a generation, is being rewritten. At last—after so much environmental damage and wasted public money—the objectives which have driven agricultural policy for the last 40 years are being re-examined.

The post-war imperative of that policy, after the years of rationing, was to increase supply. Payment was related primarily to the level of output, quality being very much a secondary consideration. This aim was pursued domestically and, when the UK joined the EC, through the Common Agricultural Policy (CAP). The latter has sought to maintain agricultural incomes almost exclusively through market support—encouraging ever greater levels of output, regardless of the consequences for our countryside, food and water quality, wildlife and animal welfare.

It is in response to those consequences that organic farming has established itself so firmly on the agricultural agenda. The production-driven CAP has become a victim of its own success. Having achieved and surpassed the production objectives which were set in the post-war period, it has paid no heed to the consequences of agricultural intensification—the 'externalities' of conventional high-input agriculture. Now that security of supply in Europe is no longer seriously in question, the emphasis must shift away from an overriding concern with quantity towards other considerations—considerations of quality.

That is why organic farming is so appropriate to the needs of the new agenda. We should recognize that agricultural policy need—and must—no longer be directed exclusively towards the production of food. It has been in the past, of necessity, as the generation which endured rationing will testify. But the success of agriculture in meeting nutritional needs has moved the debate on. We now have additional objectives: the production of food of a higher quality; and production in a way which, by considering environmental objectives, is indefinitely sustainable. We must consider simultaneously the quality of the product and the quality of the productive medium.

The status quo over the last three decades was based on the assumption that conventional farming was itself indefinitely sustainable. The first battle for the more ecologically-minded was to prove that this was not the case. Many of us have had a hunch, a gut feeling, that increasing levels of synthetic inputs could not be indefinitely absorbed by the environment; but what was required was hard scientific evidence. The increasing availability and irrefutability of that evidence has greatly strengthened the case for a reappraisal of objectives.

Secondly, if agriculture continues to be financially supported by the public—as I think in the medium term it will be, for environmental and social reasons—then it is the prerogative of the public to define those objectives. The damage to our countryside has been, and remains, a cause of great public unease. The rocketing membership of organizations involved with conservation and countryside preservation is a token of that concern. Its logical consequence is translation into formal policy objectives which ensure that environmental protection and enhancement is an explicit goal, rather than an incidental by-product, of agriculture.

If the case is made for a method of agricultural production which respects and works with, rather than against, the natural environment, two major questions remain. One is consequent upon the other. First, we must establish whether organic agriculture fits the environmental bill; and secondly, if it does, how should we build the framework within which it can flourish?

Since organic agriculture is abundantly capable of fulfilling our environmental objectives, I shall not dwell long on the first point. It is, however, true that if organic agriculture involves, for example, pest control which consists of spraying 'organic' pesticides, such as derris, rather than synthetic ones, it may be as damaging to the environment as conventional farming. This type of management emulates the conventional farmer in that it tackles the symptom rather than the cause. We must promote an organic agriculture which will be concerned with avoiding such problems through diversity, rotation and good husbandry, rather than the crisis management which characterizes so much of intensive conventional farming. Organic agriculture should be concerned more with prevention than cure.

Of course, we shall see the introduction of new technologies and new methods of husbandry which, while avoiding the use of synthetic inputs, will arguably be no more 'natural' than the conventional farmer with his crop sprayer. We shall have to develop techniques to evaluate the impact of novel organic processes, but our existing organic producers are not preoccupied with the development of ever-more sophisticated organic methods of problem solving. They're principally concerned with improving standards of management so as to avoid having to resort to such remedies—remedies which signify that the fundamental problems of intensification remain, albeit in an organic rather than conventional context. Above all, organic farming is about de-intensification, rather than finding space-age 'organic' cures for the problems of over-intensification. The organic farmer seeks to overcome problems of pests and diseases by improving standards of management, not by resorting to sophisticated technology.

It is partly because organic farming involves de-intensification that it is such an ideal part of the overall prescription to tackle the ills of the CAP. We do not want to exchange mountains of surplus conventional produce for mountains of organic produce. It is the combination of sensitivity to the environment and de-intensification—which I think are indivisible—which makes organic agriculture, as the ultimate example of less-intensive production, so well-suited to current demands.

Therein lies the belief that organic farming can help deliver the environmental benefits that we require. That being said, it is incumbent upon me, writing as a politician, to advance ideas about how organic production can be promoted.

My premise is that we cannot achieve the sort of growth in organic farming which we desire through simple reliance on the market. Admittedly, recent years have seen rapid growth without specific Government or EC incentives. However, we are unlikely to get anywhere near the hoped-for 20% by the year 2000, or substantially cut the huge proportion of demand for organic produce which is met by imports, without specific and targeted assistance. Since public finance will continue to be necessary to keep the UK agricultural industry afloat, it is surely desirable to channel that assistance towards a sector which, as I've explained, meets the new demands on agriculture.

Historically, there has been a carrot and stick effect drawing producers into the organic movement: the carrot has been the promise of premium prices for their produce, alongside a perception that organic production will in some sense be more fulfilling for them as producers. The stick has been the concern that conventional farming has been responsible for environmental degradation and expensive over-supply, neither of which is sustainable, or popular with the public.

Organic producers in the UK have supplied a highly specialized market with correspondingly high premia, upon which their financial viability has partly depended. As supply has increased, premia have already come down—though demand has grown more or less proportionately. But in the future, as the organic sector grows, the pressure on premia is likely to be in a downward direction.

However, organic farmers have not hitherto directly benefited from the fact that their methods of production have been non-polluting. Conventional farming enjoys, to a certain extent, a hidden subsidy, in that environmental costs are 'externalized', unquantified (and largely unquantifiable), and so not charged to their originator. If the principle that the polluter pays were fully implemented in agriculture, organic producers would have even lower input costs in comparison with the intensive conventional farmer than they do already, as the relative level of environmental pollution is lower.

As awareness of our environment grows—at the same time as an increasing ability to identify and measure the causes of its degradation—the use of agricultural techniques which have high environmental costs will diminish. However, there will remain an argument for assisting those methods of production which have lower environmental costs, to ensure that polluting and non-polluting techniques compete on a level playing field when all the costs of production—environmental as well as directly financial—are taken into consideration.

There are two ways of levelling the playing field. Either the polluting techniques can be taxed at the rate which is thought to recompense for the costs of their environmental damage, and thereby minimize their use; or non-polluting techniques could be financially assisted, in order directly to encourage *their* use.

In practical political terms, unless and until we have accurate methods of measuring the environmental costs of conventional agriculture, it will be simpler to assist those types of production which minimize 'external' environmental costs. There is, therefore, a strong case for assisting organic farming. It is not only desirable from an environmental point of view: if we are to take the 'polluter pays' principle seriously, it is also fairer.

This assistance should be directed at helping producers undergoing conversion from conventional to organic production. The principle of conversion grants is well established in other countries which have taken their commitment to organic farming further than the UK—such as Denmark, Sweden, Norway, Germany and Switzerland. The models vary, as does the period over which the assistance is given. The common factor is that the grant is for a specific area of conversion—usually a whole farm—and for a specific period, usually three or five years.

The risks facing those contemplating conversion are many. On the one hand, reassurance can be derived from the fact that increasing numbers of others have gone down the road of organic conversion before, which gives both confidence that it can be achieved and a pool of experience on which to draw. But on the other hand, as more producers go into organic methods, output increases, giving uncertainty about future levels of premium.

There is the additional problem of the apparent unfairness of paying new entrants to convert, without in some way giving commensurate assistance to those who have already undergone the process. After all, the pioneers have undertaken much greater risk than anyone following in their footsteps, not having enjoyed the expert advice or fund of experience now available to help newcomers get established. In addition to these existing organic producers, there are those undergoing conversion, who might therefore reasonably expect assistance even though they have decided to convert without it.

It would be foolish, however, to underestimate the political difficulties inherent in any scheme to assist financially not only those undergoing conversion, but also those who have already converted. The principle of retrospective payments would be highly contentious, particularly since many organic producers have skilfully and lucratively established niche markets. It would be more logical to approach the problem from another perspective, and establish a system of payment to farmers according to various environmental criteria.

This principle has already been established in the UK with the Environmentally Sensitive Areas, in which farmers are paid on an acreage basis to farm in a way which maintains traditional land use patterns and practices, to the benefit of the environment and landscape. The scheme has proven popular with the participants and with the public. Unlike set-aside, which has failed to deliver reduced output, ESAs have achieved their objectives.

A combination of specific payments to farmers to convert to organic production and continuing payments for environmentally-sensitive farming would encourage conversion, but also reward it where it has gone ahead unaided. Payments for sustainable management of farmland would not be available only to organic farmers. There would be a continuum of different management types, from the intensive agribusiness at one end to the fully organic mixed farm at the other, with payments on a sliding scale in between. Each producer would be perfectly free to decide where on that continuum he wishes to be; but with reducing prices for agricultural commodities, there would be a tendency to move along the continuum towards the sustainable end as real returns from the market diminished and other sources of income became increasingly necessary.

Such a system of payments would also resolve the long-term apprehension of some in the organic movement that if too many follow the organic route, the market premia which organic produce commands will be whittled away, undermining profitability and threatening viability. Most organic producers dismiss this fear, as they believe that lower input costs will increasingly tell in their favour, leaving them a continuing advantage over the conventional producer. Even if this hope proves illusory, since any system of payments made according to environmental criteria is likely to be of particular benefit to organic producers, continued viability should not be at risk. So payments should not be made retrospectively to farmers who have already undergone conversion. Instead, they would be the beneficiaries of new farm management payments made according to environmental criteria.

To tie these threads together, future support for organic producers from the political world should take the form of time-limited payments for conversion, as already happens successfully elsewhere. The formal principle of environmentally-sensitive farming should be allowed out of the ghettoes of ESAs, and payments should be made available across the country for farming in a way which benefits the environment. All farmers would stand to benefit from these, but since organic farming is inseparably linked with the quality of the environment, organic producers would clearly benefit in particular.

Policy moves away from the system of farm income being reliant solely on the level of agricultural output are already under way. The EC has proposed many measures, some mentioned above, which are intended to break the link between income and crude levels of output. The proposals I've outlined are the logical next step. Similar ideas have been advanced by the UK's farming unions, the National Farmers' Union leadership in particular. Farmers like farming, and they recognize that the price of doing so will increasingly be denominated in environmental terms. They also like a healthy environment as much as the rest of us—providing it is not at the expense of their viability. There is no earthly reason why it should be.

This way signals hope for continuing growth in organic farming, as a sector of agriculture which is singularly well-suited to the demands which we should be making of agriculture towards the end of the twentieth century. These proposals are realistic and practicable. They go with the tide of thinking in the EC, and, if implemented, they will meet with public approval, as they halt and begin to reverse the damage to our environment and countryside caused by post-war farming policies.

I think something along these lines will be implemented. I say so confident that public concern about the quality of the environment will continue to grow—and organic agriculture will continue to provide so appropriate an answer to that concern as it relates to land use and agriculture.

Chapter Nine

A Symbol of Hope—Africa's Green Belt Movement

Wangari Maathai

HEADLINES ON AFRICA IN THE WORLD PRESS tell of drought, famine, poverty, violation of human rights, debts... Only bad news from Africa hits the news-stands. In this age of the written word and the influential press, people appear to be what the press says they are. And when the people in the stories are persuaded by the press to accept the false image of themselves, it is tragic—psychologically and otherwise, for while the press cannot change reality, it can change people's perception of that reality.

Not that everything written about Africa is false. Some stories are true. There have been natural disasters, but a number of Africa's calamities have been precipitated by exploitation and by a general neglect of the continent's natural resources—her forests, wildlife, people, land, minerals and water.

There is, however, a new consciousness emerging in parts of Africa and other parts of the world. It is the awareness that every being on this planet shares a common destiny. All of Africa should be part of this awakening so that, unlike in her recent past, Africa can participate directly in the direction of her destiny. To do this, she cannot afford to ignore inspirations from her own sons and daughters.

One challenge for the future is to use our energy and develop workable strategies that will raise Africa from her economic, political and environmental crisis and propel her toward self-empowerment, self-rediscovery and self-realization. We should clearly and strongly assist African development efforts.

What is development? For many leaders and decision-makers in

Africa and throughout the world, development means extensive farming of cash crops to earn foreign exchange and huge, expensive programmes, like hydro-electric dams, airports, hospitals, luxurious tourist hotels, supermarkets, large, heavily-armed armies and imported luxury items. These are the priority items in national budget allocations.

Never mind that they may not reflect the needs of the people, who, if asked, would prefer basic needs, like food, shelter, education, clean water, local clinics, information and freedom. The people would probably prefer a different type of development—something called 'sustainable development'. Sustainable development is the utilization of national resources by the present generation in such a rational manner as would not put the future generations in jeopardy.

The Green Belt Movement is a grassroots movement with tree planting as its basic activity. Although its objectives are many and varied, it has used the tree as the focal point around which other environmental issues are discussed and brought to the attention of the public and decision-makers. Indeed, trees have become a symbol of hope. They are also a living indicator of the realization of environmental rehabilitation and conservation and of sustainable development. Tree planting itself is a simple activity. As such, it is an excellent starting point for all environmental activities.

The idea of planting trees in communities was born in 1974 as part of a plan, called Envirocare, to improve the environment of Nairobi's Langata community, but it would have died were it not for the National Council of Women of Kenya (NCWK). In 1977, I was elected to NCWK's Executive Committee and joined the Standing Committee on Environment and Habitat. After some initial opposition, the committee adopted community tree planting as a project. The first tree-planting ceremony took place in Nairobi on World Environment Day—June 5th, 1977. During the UN Conference on Desertification later that year, the NCWK decided to emphasize the practical problems of desertification facing the ordinary rural population. What had started simply as an NCWK project was gradually developing into a community activity. Interest in tree planting grew and invitations to plant trees began to trickle into the NCWK office from all over Kenya. Agroforestry has always been our way of farming and our ideas were well received. The term, Green

Belt Movement, surfaced much later with the recognized need to establish community woodlands to serve as demonstration plots. These woodlands were to be planted on the boundaries of public land so they could be easily cared for. They would also enhance the beauty of the compound and create natural windbreaks.

The objectives of the Movement were inspired by the local needs and problems of Kenya, but they apply to many other countries and communities in Africa and around the world. Some of these problems are symptoms of mistakes made in the name of 'modern development' and the mistakes are being paid for dearly—in some countries by life itself.

The Green Belt Movement endeavours to meet the needs of communities by harnessing local capabilities, expertise and resources and engaging the community to be the main driving force. The movement has deliberately discouraged direct participation by high-powered technicians and managers, shying away from imported high technology and solutions, experts and systems.

This has been done to create, encourage and promote local and indigenous expertise, abilities, ideas and resources. It has also been done to create confidence in local people, who are often overwhelmed by experts—so much so, indeed, that they begin to believe they are ignorant, inexperienced, incapable and backward, while others have solutions to all the problems. So, empowering the indigenous people is a major objective of the movement.

The Green Belt Movement attempts to address through its objectives various major needs: the need for fuelwood; the need to make tree planting an income-generating activity for women; the need to encourage self-employment; and the need to raise public awareness of environmental issues, especially among the youth. The need to prevent soil erosion is another major issue, since soil erosion is precipitated by, among other things, indiscriminate cutting down of trees and other vegetation, destruction of indigenous forests and catchment areas, cultivation along river beds, slopes and marginal lands, and irresponsible road construction. Subsistence agriculture, without the benefit of alternating crops and fields, is practised throughout poor African countries, and this also adds to soil erosion.

Despite its rapid spread, soil erosion can easily be checked by applying simple, but appropriate, preventive measures like planting the right type of trees, allowing agricultural residues to remain in the fields to decompose, maintaining vegetative cover, creating windbreaks, protecting forests and catchment areas from indiscriminate cutting, and by digging trenches, terraces and cut-off drains. When the ground is sufficiently covered by agricultural residues and other vegetation, the force of falling raindrops is broken, the speed of runoff flow is reduced and water seeps underground.

These simple and inexpensive measures can also improve soil fertility and, therefore, improve crop yield. They will also prevent the water pollution caused by eroded soil in the rivers, wells and streams from which rural people get their drinking water.

High technology is not needed. Nor experts. With a lot of willing bodies and some muscle power, it can be done. But people must understand why they, and not someone else, should do it.

We must inform and train the farmers—who are mostly women and often illiterate. Further, we need motivated and well-informed extension officers. Farmers need to realize that they have to 'feed' the soil. Since peasant farmers have always depended on shifting cultivation, it is essential that they appreciate the need to work with Mother Nature and hasten her processes of self-healing and self-rehabilitation.

To increase their crop yields, farmers need to improve the fertility or regenerative capacity of their soil with the simple and inexpensive methods available to them. For example, they can fertilize with animal manure, mulch with agricultural residues (instead of burning them), use compost and apply agroforestry principles. (Agroforestry is a method that was used traditionally, but has been ignored in favour of modern farming methods.) The practice of planting nitrogen-fixing trees with crops, for example, enhances soil fertility, retains soil moisture and reduces erosion. Hence, crop yields increase.

Food security can become a reality for Africa if Africans make agriculture a priority. Root crops and drought-resistant crops had always been Africa's hope for hard times. Coupled with improved food-processing and storage facilities, they could ease the burden of famine. These crops are ignored by Africa's elite, however, and it is these people who sit in the Ministries of Agriculture and of Economic Planning and make decisions.

Small farmers, particularly women farmers, should be supported if food production is to increase in Africa. Unfortunately, these are the farmers that are the most marginalized. They receive little tangible support and are ignored and considered peripheral in national economics. They have neither the political power nor the economic power necessary to demand better compensation for their labour or to prevent subsidized food imports. Ironically, while they have been pushed to the periphery of the economy, they are central to food production. This must change if famine is to end.

We must give priority to and empower the small farmers. Then, perhaps, agriculture in Africa will attract educated, dedicated and resourceful Africans who will be willing to produce enough food to feed Africans.

The Green Belt Movement also addresses itself to the need to save indigenous trees, shrubs and other flora. Forests, and vegetation in general, are valuable resources that provide firewood, animal fodder, fruits, honey, timber, herbal medicines, building and fencing materials. We need to prevent the extinction of our indigenous flora and fauna. We must conserve our genetic resources for their immense value in the area of medicine, food, water balance, and maintenance of the natural ecological balance.

Promotion of imported species such as eucalyptus and evergreens, purely for rapid economic returns, demonstrates short-sightedness and a certain amount of greed and selfishness. Many African wild fruits and berries, root crops and traditional grains have almost disappeared from our midst as we adopt imported crops. By doing this, we have made ourselves more willing victims of such disasters as drought.

Indiscriminate cutting of trees, regardless of how noble the reasons, jeopardizes the future and when indigenous forests and vegetative cover disappear, other parts of that biological system, like water and animal species, also disappear. Yet most people think only of plants and animals they can use or exploit today. They do not think in terms of the entire ecological system. It is time they did.

We tend to think that protecting our forests is the responsibility of the Government and the foresters. It is not. The responsibility is ours, individually. To fulfil this responsibility, we should start planting and protecting every indigenous tree and shrub within our reach—on our home compound, on our farm, in our village. There are millions of such trees and they would soon make woodlots everywhere.

The planting of trees helps in the fight against desertification. Some people imagine that desertification is an explosive phenomenon: one day, we will wake up to find the Sahara or the Kalahari desert advancing towards us like a huge flood. They also imagine that it can be stopped by a belt of trees. So they suggest that we plant trees in areas through which the desert will pass.

This is an over-simplification. Deserts will not descend on us. They are continuously being made wherever the ecological balance is upset and local biomass is reduced below a certain critical point. Once desertification begins, in our backyard for example, it propels itself and precipitates a vicious circle.

Our forefathers and mothers knew how to live with their environment. We seem to have lost that wisdom. Our ancestors offered their sacrifices to God under huge fig trees, but we have uprooted these trees to make way for tea and coffee plantations. It seems as though God created living cathedrals when He created the mighty fig trees, and guided His people to worship Him under them. But in modern times, we have built churches and cathedrals of stone and chosen to worship Him there. We have even gone further and fully destroyed God's natural cathedrals with swift axe strokes.

Is it a wonder that Mother Earth has subsequently refused to bless us with her earthly blessings? Our fields are gradually becoming barren, our springs and swamps are drying up, our livestock have started to die, and we, ourselves, are but shadows of what we used to be—God-fearing, wise, just and prosperous.

The Green Belt Movement already has many achievements to its credit. Millions of tree seedlings have been produced by women's groups, many jobs have been created, indigenous trees have been successfully introduced, school children have established public green belts, women are cultivating a more positive image of themselves, the public has become more aware of the need to protect the environment, and the Movement is spreading to other African countries. It is an example of a successful development project carried out not *for* the people, but *by* the people.

Many things have contributed to this success. In particular, the Movement has recognized the importance of identifying short-term objectives, to give its supporters immediate benefits for their efforts, though these short-term objectives must not become ends in themselves. We have been fortunate in our adequate finances and our freedom of operation, and above all in the commitment of our initiators. We have attached importance to public relations work, which has played a critical role in winning the confidence of many people from different backgrounds and of varying social and economic status. Some people feel that the Movement should become self-sustaining, for example, by selling seedlings to the public. We believe however, that the Green Belt Movement should not be expected to sustain itself. Africans still see the environment as a luxury issue for academicians and environmentalists. They have not seen the relationship between Africa's crises (i.e. widespread poverty, the food shortage and political instability) and the degraded environment. The concept of sustainable development, or development without destruction, still sounds foreign.

The Green Belt Movement hopes that Africans will, as individuals and as Governments, decide to invest in the rehabilitation and protection of the environment through community efforts like the Green Belt Movement. The very successful reforestation campaign in South Korea was largely dependent on Government funds, incentives for the peasants and rapid electrification of the rural areas to cut down on the use of fuelwood. This and other projects have shown that people must be willing to spend their tax money on reforestation, reclamation, and rehabilitation of the environment in order for them to work.

Community reforestation projects can be carried out cheaply—especially if supplemented with human resources, and Africans ought to understand their problems well enough to utilize all available resources. Government and donor agencies should be willing to support such efforts with money, since they are economically viable, that is, they offer employment and ensure sustainable development. While projects like these will not generate income enough to please classical economists, they can be seen as long-term investments of national importance. They are economics 'as if people mattered'.

People often complain that they don't have enough land. However, each family plot or compound has a higher capacity for trees than most people think. Trees can be planted along the compound's boundaries and along walkways and highways. Crops can be intercropped with fruit trees. By applying more efficient management, most small-scale farmers can plant many more trees on their land before reaching the maximum carrying capacity for trees.

People who live in urban areas or who have no land of their own should also plant trees, as well as shrubs and flowers, on the compounds where they live. Wherever one lives, even if only for a day, is one's home. Throughout the country, there are pieces of land lying idle. They should be planted with more trees, shrubs and food crops to improve the general environment and to enhance food production. Planting even a single tree or flower is a good start. Thanks to those who went before us, many of us today enjoy a beautiful environment which provides shade, windbreaks, birds, butterflies and other wonders of creation. What shall we, in turn, leave for others? If everyone plays their part and cares for the environment of our countryside, towns and cities, we will one day live in the type of country which most of us wish for but are not willing to work for.

In our 10 years of field experience in Kenya, we have developed a procedure for spreading the conservation message that has produced good results. The elements of this procedure may also work in other developing countries.

The message should respond to a local need, and make good common sense. Any project which is established should be honestly run and designed specifically to benefit a community, and it should work patiently to motivate that community. In taking the message to rural communities, the teachers must also become the pupils. We all, whether leading a project or participating in one, have much to learn from each other. The conservation message should start by providing basic needs and slowly become personalized so that each member of the community accepts it. The dialogue must continue, therefore, until we all believe that the protection of our world is for the benefit of all.

The wider, political dimensions must not be ignored. Politicians and decision-makers in the developing world are the rich, the elite and the powerful. Their lifestyle is expensive and in many instances the plundering of the environment is being done either by them or for them. Many pay only lip service to conservation. However, it is impossible to take the message of conservation to the rural communities effectively without the support of the decision-makers. They must realize that it is to their benefit if the masses work to prevent desertification. The message should be taken to the powerful and to the communities almost simultaneously. Even though it may take the decision-makers a long time before their support is more than rhetoric, a verbal commitment to the project is essential. The rural community, too, will be even more enthusiastic about the project if their leaders are supportive.

Only time will tell if the Green Belt Movement has made a difference, but while so many dark images of Africa are presented to the world, while development strategies, national organizations and leaders are presented as inadequate, the Movement is at the least a positive effort coming from the people to affect their world. The Movement will continue its activities in pursuit of both local and regional objectives. It will follow the same strategy, which is continuously adapting and evolving with our experiences. We know that while the results to date are impressive, the challenges are still enormous.

The healing of Africa is still only a dream. Like others before us who have faced huge challenges, however, we believe that we shall overcome.

Chapter Ten

Sustainable Agriculture in the USA

John P. Reganold, Robert I. Papendick
and James F. Parr

For nearly four decades after World War Two, US agriculture was the envy of the world, almost annually setting new records in crop production and labour efficiency. During this period US farms became highly mechanized and specialized, as well as heavily dependent on fossil fuels, borrowed capital and chemical fertilizers and pesticides. Today the same farms are associated with declining soil productivity, deteriorating environmental quality, reduced profitability and threats to human and animal health.

A growing cross-section of American society is questioning the environmental, economic and social impacts of conventional agriculture. Consequently, many individuals are seeking alternative practices that would make agriculture more sustainable.

Sustainable agriculture embraces several variants of non-conventional agriculture that are often called organic, alternative, regenerative, ecological or low-input. Just because a farm is organic or alternative does not mean that it is sustainable, however. For a farm to be sustainable, it must produce adequate amounts of high-quality food, protect its resources and be both environmentally safe and profitable. Instead of depending on purchased materials such as fertilizers, a sustainable farm relies as much as possible on beneficial natural processes and renewable resources drawn from the farm itself. Sustainable agriculture addresses many serious problems afflicting US and world food production: high energy costs, groundwater contamination, soil erosion, loss of productivity, depletion of fossil resources, low farm incomes and risks to human health and wildlife habitats. It is not so much a specific farming strategy as it is a system-level approach to understanding the complex interactions within agricultural ecologies.

In 1980 the US Department of Agriculture (USDA) estimated between 20,000 and 30,000 farmers—about one per cent of the nation's total—were practising non-conventional agriculture, most of which could now be termed sustainable. Today some experts estimate that the figure may have doubled or tripled. Farmers who practise soil conservation and reduce their dependence on fertilizers and pesticides generally report that their production costs are lower than those of nearby conventional farms. Sometimes the yields from sustainable farms are somewhat lower than those from conventional farms, but they are frequently offset by lower production costs, which leads to equal or greater net returns.

To understand the rationale for sustainable agriculture, one must grasp the critical importance of soil. Soil is not just another instrument of crop production, like pesticides, fertilizers or tractors. Rather it is a complex, living, fragile medium that must be protected and nurtured to ensure its long-term productivity and stability.

Healthy soil is a hospitable world for growth. Air circulates through it freely, and it retains moisture long after rain. A tablespoon of soil contains millions of grains of sand, silt and clay and has a vast expanse of internal surface area to which plant nutrients may cling. That same tablespoon of soil also contains billions of micro-organisms, including bacteria, actinomycetes, fungi and algae, most of which are principal decomposers of organic matter. Decomposition results in the formation of humus and the release of many plant nutrients. The microbes also produce sticky substances called polysaccharides that glue soil particles together and help the soil to resist erosion.

Another essential activity that takes place in the soil is the fixation of nitrogen. Certain bacteria in the soil or in the roots of plants (most notably legumes) convert atmospheric nitrogen gas into fixed forms of nitrogen that plants and other organisms use to make proteins. The amount of available nitrogen strongly influences soil productivity.

One of the earliest landmarks of the sustainability movement in the US is *Farmers of Forty Centuries: Permanent Agriculture in China, Korea and Japan*, by Franklin King, published in 1911. It documents how farmers in parts of East Asia worked fields for 4,000 years without depleting the fertility of their soil. This text and others of the early twentieth century focused on holistic aspects of agriculture and the complex interactions within farming systems.

Yet around this same time, US agriculture was in the early stages of industrialization. New technologies and scientific methods were developed to help farmers meet the growing demands of expanding urban populations. By substituting mechanical power for horses, for example, farmers could increase their grain acreage by from 20 to 30 per cent, because they could plough more ground in less time and did not need to grow fodder.

Many groups and individuals continued to believe that biology and ecology rather than chemistry and technology should govern agriculture. Their efforts helped to give birth to the soil conservation movement of the 1930s, the ongoing organic farming movement and considerable related research. Nevertheless, by the 1950s technological advances had caused a shift in mainstream agriculture, creating a system that relied on agrichemicals, new varieties of crops and labour-saving, energy-intensive farm machinery. This system has come to be known as conventional farming.

As pesticides, inexpensive fertilizers and high-yielding varieties of crops were introduced, it became possible to grow a crop on the same field year after year—a practice called monocropping—without depleting nitrogen reserves in the soil or causing serious pest problems. Farmers began to concentrate their efforts on fewer crops. Government programmes promoted monoculture by subsidizing only the production of wheat, corn and a few other major grains. Unfortunately, these practices set the stage for extensive soil erosion and for pollution of water by agrichemicals.

In the US between 1950 and 1985, as a share of total production cost, the cost of interest, capital-related expenses and manufactured farm inputs (such as chemical fertilizers, pesticides and equipment) almost doubled from 22 to 42 per cent, while labour and on-farm input expenses declined from 52 to 34 per cent. During most of this period, relatively little research on sustainable agriculture was conducted because of lack of funding and public interest.

By the late 1970s, however, concerns were mounting that rapidly rising costs were endangering farmers nationwide. In response, Secretary of Agriculture Robert S. Bergland commissioned a study in 1979 to assess the extent of organic farming in the US, as well as the technology behind the farming and its economic and ecological impact. The study, *Report and Recommendations on Organic Farming*, published in 1980, was based heavily on case studies of 69 organic farms in 23 states.

The USDA report concluded that organic farming is energy-efficient, environmentally sound, productive and stable and tends toward long-term sustainability. Since the report was published, it has stimulated interest, nationally and internationally, in sustainable agriculture. Its recommendations provided the basis for the alternative-agriculture initiative passed by Congress in the Food Security Act of 1985, which calls for research and education on sustainable farming systems.

The sustainable agriculture movement received a further boost in September 1989 when the Board on Agriculture of the National Research Council released another study, *Alternative Agriculture.* Although controversial, the report is perhaps the most important confirmation of the success of farms that rely on biological resources and their beneficial interactions instead of chemicals. It found that well- managed farms growing diverse crops with little or no chemicals are as productive and often more profitable than conventional farms. It also asserted that 'wider adoption of proven alternative systems would result in even greater economic benefits to farmers and environmental gains for the nation'.

Sustainable agriculture does not represent a return to pre-industrial revolution methods; rather it combines traditional conservation-minded farming techniques with modern technologies. Sustainable systems use modern equipment, certified seed, soil and water conservation practices and the latest innovations in feeding and handling livestock. Emphasis is placed on rotating crops, building up soil, diversifying crops and livestock and controlling pests naturally.

Whenever possible, external resources—such as commercially purchased chemicals and fuels—are replaced by resources found on or near the farm. These internal resources include solar or wind energy, biological pest controls and biologically fixed nitrogen and other nutrients released from organic matter or from soil reserves. In some cases external resources may be essential for reaching sustainability. As a result, such farming systems can differ considerably from one another because each tailors its practices to meet specific environmental and economic needs.

A central component of almost all sustainable farming systems is the rotation of crops—a planned succession of various crops growing on one field. When crops are rotated, the yields are usually 10 to 15 per cent higher than when they grow in monoculture. In most cases monocultures can be perpetuated only by adding large amounts of fertilizer and pesticide. Rotating crops provides better weed and insect control, less disease build-up, more efficient nutrient cycling and other benefits.

A typical seven-season rotation might involve three seasons of planting alfalfa and ploughing it back into the soil, followed by four seasons of harvested crops: one of wheat, then one of soybeans, then another of wheat and finally one of oats. The cycle would then start over. The first season of wheat growth would remove some of the nitrogen produced by the alfalfa; the soil's nitrogen reserves would be depleted much less by the soybeans, which are legumes. Oats are grown at the end of the cycle because they have a smaller nutrient requirement than wheat.

Regularly adding crop residues, manures and other organic materials to the soil is another central feature of sustainable farming. Organic matter improves soil structure, increases its water-storage capacity, enhances fertility and promotes the tilth, or physical condition, of the soil. The better the tilth, the more easily the soil can be tilled and the easier it is for seedlings to emerge and for roots to extend downward. Water readily infiltrates soils with good tilth, thereby minimizing surface runoff and soil erosion. Organic materials also feed earthworms and soil microbes.

The main sources of plant nutrients in sustainable farming systems are animal and green manures. A green manure crop is a grass or legume that is ploughed into the soil or surface-mulched at the end of a growing season to enhance soil productivity and tilth. Green manures help to control weeds, insect pests and soil erosion, while also providing forage for livestock and cover for wildlife.

By raising a diverse assortment of crops and livestock, a farm can buffer itself against economic and biological risks. Diversity results from mixing species and varieties of crops and from systematically integrating crops, trees and livestock. When most of North Dakota experienced a severe drought during the 1988 growing season, for example, many monocropping wheat farmers had no grain to harvest. Farmers with more diversified systems, however, had sales of their livestock to fall back on or were able to harvest their late-seeded crops or drought-tolerant varieties. A biologically diverse farming system is also less susceptible to the economic woes of a flooded market or a fall in prices for a single crop.

Controlling insects, diseases and weeds without chemicals is also a goal of sustainable strategies, and the evidence for its feasibility is encouraging. One broad approach to limiting use of pesticides is commonly called integrated pest management (IPM), which may involve disease-resistant crop varieties and biological controls (such as natural predators or parasites that keep pest populations below injurious levels). Farmers can also select tillage methods, planting times, crop rotations and plant-residue management practices to optimize the environment for beneficial insects that control pest species or to deprive pests of a habitat. If pesticides are used as a last resort, they are applied when pests are most vulnerable or when any beneficial species and natural predators are least likely to be harmed.

In practice, IPM programmes are a mixed bag. They have dramatically reduced use of pesticides on crops such as cotton, sorghum and peanuts. More than 30 million acres (about eight per cent of US farmland) is currently being managed with IPM programmes, resulting in annual net benefits of more than $500 million. On the other hand, IPM programmes have also been reduced to 'pesticide management' for many crops like corn and soybeans, for which pesticide usage has actually increased significantly.

Biological-control techniques are some of the best ways to control pests without pesticides. They have been used for more than 100 years and have been commercially successful in controlling pests, especially insects, in more than 250 projects around the world. Yet USDA funds for studying them have declined.

Can sustainable farming practices make good on their promise to be ecologically stable, productive and profitable? To compare the effects of sustainable and conventional farming systems on soil productivity, one of us (Reganold), working with Lloyd F. Elliott and Yvonne L. Unger of Washington State University, conducted a study of the soil on two commercial wheat farms. One was an 800-acre sustainable farm that had been managed without synthetic fertilizers and with only limited amounts of pesticides since it was first ploughed in 1909. The other was an adjacent 1,300-acre conventional farm, which had first been cultivated in 1908 but had been treated with fertilizers since 1948 and with pesticides since the early 1950s. The sustainable farm used a complex crop rotation system and practised conservation-oriented methods of tillage, whereas the conventional farm followed a simple two-year rotation. The sustainable farm also grew legumes as a cover crop and green manure.

Because of the differences in farming methods, the soil on the sustainable farm contained significantly more organic matter, nitrogen and biologically available potassium than that on the conventional farm. It had a better capacity for storing nutrients, a higher water content, a larger micro-organism population and a greater polysaccharide content. The soil also had better structure and tilth and 16 more centimetres of crop-nourishing topsoil. This topsoil difference was attributed to significantly more soil erosion on the conventional farm.

Average yields of winter wheat per acre between 1982 and 1986 were eight per cent lower on the sustainable farm than on the conventional farm. Nevertheless, the sustainable farm matched the wheat production average for the region—in fact, it yielded almost 13 per cent more wheat than another nearby conventional farm with similar soils. Its ability to do so, even after almost 80 years of farming without fertilizer, may result in part from reduced soil erosion and maintenance of soil productivity.

Although conserving soil productivity is important to farmers, most of them usually select an agricultural system on the basis of its short-term profitability. Until recently, conventional systems have usually appeared to be more profitable in the short term than sustainable ones. This assessment comes as no surprise, because research and USDA policy over the past four decades have promoted conventional agriculture.

Yet the long-term profitability of conventional agriculture seems questionable if the environmental and health costs currently borne by society are taken into account. If these indirect costs were factored into the costs of conventional farm production, then sustainable systems would likely prove to be more profitable and more beneficial to society.

One of the best-known studies of the economics of sustainable agriculture was conducted by William Lockeretz, Georgia Shearer and Daniel H. Kohl of Washington University. They compared energy efficiency and crop production costs between numerous pairs of organic and conventional farms in the Midwest. Between 1974 and 1978 the energy consumed to produce a dollar's worth of crop on the organic farms was only about 40 per cent as great as on the conventional farms. Although the organic farms had lower crop yields than the conventional farms, their operating costs were lower by about the same cash equivalent. As a result, the net incomes from crop production on the two types of farms were about equal every year except one.

Despite these encouraging results, some farmers who have shifted from conventional to sustainable practices have experienced short-term difficulties. Some of the problems arose because the farmers abruptly stopped applying pesticide and fertilizer to all their fields. Such radical changes can sometimes decrease yields because of severe weed problems, explosive increases in insect pests and diminished soil fertility that lasts a few years.

Researchers at the Rodale Research Center in Kutztown, Pa., have investigated the transition from conventional to sustainable farming and verified that such changes are best implemented with caution. Even a gradual change may involve small decreases in crop yields while the soil establishes a new set of chemical and biological equilibria. Farmers should change only one field at a time to avoid placing whole farms at risk. The transition is also smoother if they regularly add organic matter to the soil in the form of animal or green manures.

What are the forces that inhibit farmers from adopting sustainable methods? One obstacle is the federal farm programmes, which generally support prices for only a handful of crops. Corn and other feed grains, wheat, cotton and soybeans receive roughly three-fourths of all US crop subsidies and account for approximately two-thirds of cropland use. The lack of price supports for other crops effectively discourages farmers from diversifying and rotating their crops and from planting green manures. Instead it gives them powerful incentive to practise monoculture to achieve maximum yields and profits.

The long-term economic benefits of sustainable agriculture may not be evident to a farmer faced with having to meet payments on annual production loans. Many conventional farmers are greatly in debt, partly because of heavy investments in specialized machinery and other equipment, and their debt constrains the shift to more sustainable methods. To date, society has neither rewarded farmers financially nor given them other incentives for choosing sustainable methods that would benefit the public.

Then, too, there is little information available to farmers on sustainable practices. Government-sponsored research has inadequately explored alternative farming and focused instead on agrichemically-based production methods. Agribusinesses also greatly influence research by providing grants to universities to develop chemical-intensive technologies for perpetuating grain monocultures.

Legislative support for change in the US agricultural system is growing, but financial support for sustainable agricultural projects is still only a small part of the total outlay for agriculture. Congress appropriated $3.9 million in fiscal year 1988 and $4.45 million in fiscal 1989 to implement the research and education programmes on sustainable farming called for in the Agricultural Productivity Act, one part of the Food Security Act of 1985. Funding in fiscal 1990 was the same as in the previous year—$4.45 million—which is only 0.5 per cent of the total USDA research and education budget.

The programme for low-input sustainable agriculture, or LISA, that has emerged from this federal effort has many objectives: to reduce reliance on fertilizer, pesticide and other purchased resources to farms; to increase farm profits and agricultural productivity; to conserve energy and natural resources; to reduce soil erosion and the loss of nutrients; and to develop sustainable farming systems.

A 1988 US House of Representatives report, *Low Input Farming Systems: Benefits and Barriers*, recommended that Congress restructure or remove some provisions in farm-support programmes, particularly those that encourage greater use of agrichemicals and that impede the adoption of low-input methods. In 1989 three congressional bills were introduced—two in the Senate and one in the House of Representatives—that would allow farmers to rotate crops and use other alternative methods without losing farm-support funds.

Shifting mainstream agriculture towards more sustainable methods will require more than new laws and regulations; it will also require more research and public education. Universities and the USDA are slowly putting more emphasis on sustainable agricultural research. A high research priority is the development of specific cropping systems that produce and consume nitrogen more efficiently. It is essential to learn how much nitrogen is fixed by legumes under various conditions, as well as the optimum means for integrating legumes into crop rotations.

The US should also step up its research efforts on other topics. More must be learned about alternatives to fertilizers and the cycling of nutrients through the agricultural ecosystem. Effective strategies must be developed for controlling pests, weeds and diseases biologically. The strategies may rely on beneficial insects and microorganisms, allelopathic crop combinations (which discourage weed growth), diverse crop mixtures and rotations and genetically resistant crops. More research should also be done on the relative benefits of various cover crops and tillage practices and on integrating livestock into the cropping system.

US farmers now use only a fraction of the thousands of crop species in existence. They may benefit by increasing cultivation of alternative crops such as triticale, amaranth, ginseng and lupine, which are grown in other countries. Yet in addition to diversification, germ plasm (seeds, root stocks and pollen) from traditional crops and their wild relatives must be collected and preserved continually.

Well-managed collections of germ plasm will give plant breeders a broader genetic base for producing new crops with greater resistance to pests, diseases and drought. Today much of the germ plasm that US plant breeders use to improve crops comes from developing countries.

New breeds of crops being developed by biotechnology, such as grains that fix their own nitrogen, may eventually be included in sustainable cropping systems. But neither biotechnology nor any other single technology can fix all the problems addressed by a balanced ecological approach. The success of sustainable agriculture does not hinge on creating supercrops: the system works with crops that are available now.

Better education is as important as further research. Farmers need to know clearly what sustainable agriculture means, and they must see proof of its profitability. The USDA and the Co-operative Extension Service should provide farmers with information that is up-to-date, accurate, practical and applicable to local farming conditions. Farmers and the public also need to be better educated about the potentially adverse environmental and health consequences of the pollution created by certain agrichemical practices.

One of the most effective methods for communicating practical information about sustainable agriculture is through farmer-to-farmer networks, such as the Practical Farmers of Iowa. Farmers in this association have agreed to research and demonstrate sustainable techniques on their lands. They meet regularly to share information and compare results. Because such networks have aroused growing interest and proved effective, the land-grant community should try to promote their development.

Some scientists and environmentalists have recommended levying taxes on fertilizers and pesticides to offset the environmental costs of agrichemical use, to fund sustainable agricultural research and to encourage farmers to reduce excessive use of agrichemicals. This approach is precisely how funding for the Leopold Center for Sustainable Agriculture was established by the Iowa State Legislature in 1987 as part of the Iowa Groundwater Protection Act.

Agriculture is a fundamental component of the natural resources on which rests not only the quality of human life but also its very existence. If efforts to create a sustainable agriculture are successful, farmers will profit and society in general will benefit in many ways. More important, the US will protect its natural resources and move closer towards attaining a sustainable society.

Chapter Eleven

Towards the Marriage of Ecology and Economics

Wes Jackson

WHERE NATURE IS AT WORK there is an ecological mosaic, a world of diverse ecosystems. How to handle this mosaic between and including extremes should determine our agenda for the next century if sustainability is our watchword. A tension arises because we are Homo, the homogenizer with a generalizing technological capability. Whether it is a Nebraska prairie community or the tropical rainforest or anything else in between, problems arise because we humans tend to invert what nature does well. If nature is our standard or measure for a sustainable culture, then we should expect a mosaic of human cultures or communities across our land. Our problem then becomes, how much of this mosaic do we dare aggregate if we are to do it safely and sustainably? At a recent conference held at the University of Minnesota, a soil expert cited a 54-acre field which had seven mappable units, which is not uncommon especially for the glaciated agricultural soils of North America. For corn alone, the yield potential of these seven units varied from 112 to 162 bushels per acre. A fertilizer expert could logically recommend anywhere from 88 to 132 pounds of nitrogen per acre per year for that 54-acre field, depending on the unit. The phosphorus need, based on the soil tests, ranged from 10 to 35 pounds of P_2O_5 per acre. Here is why it is important to think of small scale, or a high eyes-to-acres ratio in a world of scarce resources. Farmers usually apply a fertilizer level their *best* soils can handle which leads to a waste of the resource and can cause pollution of the groundwater. These seven soil types are part of that ecological mosaic I mentioned, the mosaic for which we must provide the 'elegant solutions', the mosaic which becomes the basis for countless mini-economies.

It may appear that my proposed solution, the only solution that is unambiguously ecologically correct, is a cultural solution; one which rewards small scale, but here's the rub. This very field was described to illustrate the fact that information from soil surveys and production data is becoming available on floppy disks. This new technology can accommodate a new, but still infant, fertilizer spreader equipped with a microprocessor designed to control the appropriate distribution of fertilizer as the machine proceeds across a field. This machine is predicted to have a major future role in reducing pollution. In other words, even here, where cultural intelligence could play a major role and justify small farm size, technological cleverness is introduced to take care of what is regarded as the main problem—pollution. This is why something which is conceptually rich requires cultural intelligence which must override any technological 'tour de force', or John Todd's 'elegant solutions being predicated on the uniqueness of place' will become simply the battlecry for an extension of the old industrial mind into a new technological frontier. It is a perception of the world that is on the line.

The question now becomes: where do we look for the primordial roots of that necessary perception? Vaclav Havel, Czechoslovakia's leading playwright and new president, provides an insight in a wonderful essay entitled 'Politics and Conscience'. He begins by relating how:

> As a boy, I lived for some time in the country and I clearly remember an experience from those days: I used to walk to school in a nearby village along a cart track through the fields and, on the way, see on the horizon a huge smokestack of some hurriedly-built factory, in all likelihood in the service of war. It spewed dense brown smoke and scattered it across the sky. Each time I saw it, I had an intense sense of something profoundly wrong, of humans soiling the heavens. I have no idea whether there was something like a science of ecology in those days; if there was, I certainly knew nothing of it. Still that 'soiling the heavens' offended me spontaneously. It seemed to me that, in it, humans are guilty of something, that they destroy something important, arbitrarily

> disrupting the natural order of things, and that such things cannot go unpunished. To be sure, my revulsion was largely aesthetic; I knew nothing then of the noxious emissions which would one day devastate our forests, exterminate game and endanger the health of people.[1]

Havel describes this feeling he had as a primordial response, a response whose *source* is the natural world, a world to which we are personally bound 'in our love, hatred, respect, contempt, tradition, in our interests'. He believes that out of this world 'culture is born' and that there is a moral order in this natural world in the sense that:

> Any attempt to spurn it, master it or replace it with something else, appears, within the framework of the natural world, as an expression of *hubris* for which humans must pay a heavy price, as did Don Juan and Faust.

[Returning to the smokestack, he elaborates by saying:]

> To me, personally, the smokestack soiling the heavens is not just a regrettable lapse of a technology that failed to include the 'ecological factor' in its calculation, one which can be easily corrected with the appropriate filter. To me it is more, the symbol of an age which seeks to transcend the boundaries of the natural world and its norms and to make it into a merely private concern.[2]

And later:

> A modern man, whose natural world has been properly conquered by science and technology, objects to the smoke from the smokestack only if the stench penetrates his apartment. In no case, though, does he take offence at it *metaphysically* since he knows that the factory to which the smokestack belongs manufactures things that he needs. As a man of the technological era, he can conceive of a remedy only within the limits of technology—say, a catalytic scrubber fitted to the chimney.[3]

Havel is not out to abolish smokestacks or prohibit science or go back to the Middle Ages. He is trying to find the tentacles of the monster we call hubris, the assumption that our knowledge is adequate to run the world.

Reading his essay, I was reminded of Descartes' *Discourse on Method,* in which he said that the more he sought to inform himself the more he realized how ignorant he was. But rather than regard this as an apt description of the human condition and the very proper result of a good education, Descartes saw our ignorance as correctable even though, when we think about it, we are billions of times more ignorant than knowledgeable.

We are billions of times more ignorant, for example, about the area affected by Chernobyl—which reaches all the way to the Antarctic now—than we are knowledgeable and our ability to correct the consequences of that one accident is clearly zero.

Havel is refreshing because, even though his essay is about politics, his grounding is ecological, at least agro-ecological. Hear this example:

> For centuries, the basic component of European agriculture had been the family farm. In Czech, the older term for it was *grunt*—which itself is not without its etymological interest. The word, taken from the German *Grund,* actually means ground or foundation and, in Czech, acquired a peculiar semantic colouring. As the colloquial synonym for 'foundation', it points out the 'groundedness' of the ground, its indubitable, traditional and pre-speculatively given authenticity and veridicality. Certainly, the family farm was a source of endless and intensifying social conflict of all kinds. Still, we cannot deny it one thing: it was rooted in the nature of its place, appropriate, harmonious, personally tested by generations of farmers and certified by the results of their husbandry. It also displayed a kind of optimal mutual proportionality in extent and kind of all that belonged to it; fields, meadows, boundaries, woods, cattle, domestic animals, water, toads and so on. For centuries no farmer made it the topic of a scientific study. Nevertheless, it constituted a generally satisfactory economic and ecological system, within which everything was bound together by a thousand threads of mutual and meaningful

> connection, guaranteeing its stability as well as the stability of the product of the farmer's husbandry. Unlike present-day 'agrobusiness' the traditional family farm was energetically self-sufficient. Though it was subject to common calamities, it was not guilty of them—unfavourable weather, cattle disease, wars and other catastrophes lay outside the farmer's province.[4]

I suspect there was soil erosion on those Czech farms and that some catastrophes *were* within the farmer's province. Nevertheless, Havel's political views are rooted in the fact that we are land animals, not industrial or space colony people.

Our work at The Land Institute involves a marriage of ecology and agriculture—agroecology it is called. Our approach is to *mimic* the native prairie, which features perennials grown in a mixture. We clip, sort and weigh the above-ground plant material to establish the ratio of grasses to legumes to sunflowers to others in order to mimic that vegetative structure with the perennials we are breeding. Nature is the standard or measure. We, too, want to conform to nature's economy as much as is practical. But here's the problem. We may be scientists, but it would be naive for us to suppose we can create a satellite of agricultural sustainability which can safely orbit a culture featuring an extractive economy. Therefore, our work has to move beyond a marriage of ecology and agriculture to include a marriage of ecology and economics. Capitalism is not grounded in ecological necessity for, as Herman Daly says, it more nearly resembles the ideology or economics of the digestive tract than the circulatory system. Whatever the metaphor, this product of the industrial mind, our economic system called capitalism, is bankrupting America. It was designed to accommodate the exploitation of the hemisphere and became really big during this fossil fuel epoch. The ecological debt is now being translated into a dollar debt.

Just as Havel grounds his politics in nature, what if we were to ask 'What is nature's economy?' What kind of an economy does the prairie have and, for that matter, all the other ecosystems before humans make their break with nature through agriculture? For millions of years, nature's economy has handled the cycling of materials essentially all on solar energy. What can we learn about how nature pays the bills? One thing we can be sure of as we look to nature and that is that she has no comfortable absolutes to which we can attach ourselves. Nature both mines and pollutes. Nosing roots will mine nutrients from parent rock material and build top soil. Fossil fuel deposits, after all, represent a recycling failure and contributed to the build-up of atmospheric oxygen which initially was a noxious environment, but, eventually it led to the development of aerobic organisms.

No philosopher spends much time shopping for absolutes any more. Nevertheless, nature is still a good standard, because most failures have been slow—spread over many millions of years. In a larger sense, what has characterized nature's ecosystems is an *economy* of nutrient recycling and a graceful handling of the solar flux for all kinds of work, including the support of life. That is why those involved in ecological restoration are engaged in re-establishing a renewable economy, not economic renewal. It is religious work in the sense that restoration has to do with redemption, a religious term which probably has an ecological source. But here again, in an absolute sense, the world doesn't have to be redeemed by people. Nature has the power to redeem itself and did so long before we arrived. The Amazon carries a huge supply of nutrients towards the sea each second. It once ran the other way, towards the Pacific. Thanks to a great work of nature, the Andean uplift, the direction was reversed. Had I been alive at the time all these nutrients were headed towards the Pacific for burial and had agriculture or deforestation been a major contributor, I would have campaigned against our contribution to this loss, because, in our human time frame, *potential* was being destroyed. The uplift which sent the river back in the other direction is in geological, not human time. We need to be clear on what kind of time period we're talking about, for this mixing of time frames has made it possible to excuse our destructive behaviour on the grounds that nature does it, too.

As long as we are within human time, then our actions have to be scrutinized. Much of the language the ancients employed in the development of a moral code is a recognition of ecological necessity within human time.

I mentioned earlier that redemption is an ecological term. An eroded hillside, with loving care, can be redeemed, perhaps within two or three generations. That is well within human time. It is when we move beyond generation time, beyond historical and even beyond prehistorical time that our moral code can become confused. It wasn't until we started agriculture that our impact on the geological scale of time developed, with soil erosion beyond replacement levels.* Since then we have probably added global warming and certainly acid rain and the ozone hole. But what do we do if we don't know whether indeed our actions are damaging? Mentioning global warming brings to mind another question. A growing number of scientists believe it to be responsible for an impending, if not already present, sea-level rise, but what do we do if we don't really know whether the carbon accumulating in the atmosphere is causing the world to get warmer? Earth scientists are having trouble deciding on the most reliable places to place their instruments. Sea level is not a constant. The entire north-eastern part of the United States is still rebounding from the last glacier, which retreated some 15,000 years ago, which means that New England rebounds while the south-eastern United States subsides. The Pacific is higher than the Atlantic, because as the Earth turns, water piles up on the west side of continents. We don't know what's happening to polar ice; it might actually be accumulating as the globe warms. On the other hand, there is a seaward slip of Antarctic ice that, if it continues, could raise the ocean by up to 20 feet. What is sea level anyway?

I mention these sorts of massive dynamics, some of which involve long periods of time and space, partly because one wonders to what extent any moral law can be derived from nature. We need to consider whether moral laws can be independent of natural laws. In the meantime there are some basics we can explore. We need to come to terms with the very basis of our knowledge in the modern world of science. To that end, we are forced back to the early 1600s when Francis Bacon explicitly said that we must bend nature to our will. René Descartes added that to do so, we must place priority on the parts of things over the whole. The worst assumption of the modern scientist is that the world is like this method. Furthermore, this method for dominion and power over nature does not acknowledge the fact that, as the world is broken down, the emergent qualities are lost.

*The extinction of the megafauna by American Indians may be an exception.

Nevertheless, that became the modern world view on how we come to understand the world. This is the beginning of modern science and remains the dominating lens through which scientists work.

How do we cope with the fact that we are, on occasion at least, prone to ignorance and wickedness? Simply this: when we set out to rearrange the world, the best way to minimize destruction is to keep the scale small, so that we are able to cut the losses before they balloon.

I have given examples of how nature both mines and pollutes, but my larger point is that the amount of time is more often geological than what I described as human time.

I have also spoken against a knowledge-based world view as a most destructive assumption of Descartes. Descartes is right, however, if our scale is small enough. Fortunately, this helps us ponder what constitutes good judgement as to where we fit in a time perspective as well as in space. I come back to what is featured in nature's economy, like a tallgrass prairie ecosystem. What we see is a *community* of life forms, a diversity of *species*. A human-dominated ecosystem also features community, but with a diversity of *persons*. It is interesting that the locus for both is community, community ecology.

The marriage of ecology and agriculture has produced a hybrid called agroecology. Since the break with nature came with agriculture, it is fitting that agroecology arrange a long-term courtship between ecology and economics.

Before there can be a marriage, the current perception of the run-of-the-mill economist, like that of the run-of-the-mill agriculturalist, will have to be swept into the dustbin of history. But perceptions can change. A small amount of progress has already been made by a few agriculturalists. A few agricultural researchers are now willing to consider where our domesticated livestock spent most of their evolutionary history. By acknowledging that our domesticated plants and animals are relatives of wild things which had a long evolution in a context that is not of our making, more agricultural researchers in the future should be able to think of the chicken as a jungle fowl rather than primarily as property we have a licence to confine in small cages to produce eggs or meat. The hog is a forest animal and, therefore, not just a pin cushion for needles which squirt the antibiotics close confinement requires. Both the beef and the milk cow had their origin as grazers of savannah-like conditions, not feedlot critters or milk machines designed to eat like a hay baler.

Because our reference point at The Land Institute is nature's prairie, the application of our technological array and scientific know-how is subordinate to a standard which features neither efficiency nor production. It is not that we are uninterested in efficiency or production, and as we go about our daily work a visitor may not be able to see our motions as different from the motions and patterns of other agricultural researchers. We work in a fancy greenhouse under careful temperature control in late fall, all winter long and into the spring. Here seeds germinate and grow and numerous pollinations are made. We work in orderly experimental plots in the field. But as the breeder pollinates, the pathologist evaluates damage and the ecologist dopes out certain soil-root interrelationships, and standing firmly in the background is that never-ploughed native prairie whose living community is under continuing scrutiny. The products of that mental shift can't be readily seen, not yet.

I could restrict the argument to life forms, but the fact is it isn't just the living world we are talking about here in this mental shift. It is all of nature. Think about water in terms of a proper relationship, for example. With water, we can apply the same principle which guides our thinking on how to treat our livestock. It is the ecological context which remains foremost. It is one thing for a farmer adjacent to a stream to divert a small quantity of water to irrigate a few acres and quite another for a stream to become acre-feet delivered in a 400-mile-long corridor of pipes and ditches to Los Angeles. The question is whether the stream, as an ecosystem or part of one, can experience or 'enjoy' streamhood. It is analogous to a few trees being removed from a woodlot to become useful lumber rather than clear cutting, which turns the woodlot into board-feet—a commodity. Acre-feet and board-feet are the same—resources.

The ecological perspective, in other words, honours *jungle* fowl, *forest* animals and *savannah* grazers—and in the future, I hope, domestic *prairie* seeds. I emphasise the adjectives because the adjective in these four cases is the ecosystem which describes the *relationship* of the larger system to the creature. The ecological perspective honours the woods and the stream, the tropical rainforest, *and the Kansas prairie.* As we move towards the larger notion of relationship or context, the word 'resources' becomes obscene.

I seriously doubt that the US Department of Agriculture or the National Academy of Sciences or the US Congress will one day soon move from the notion of 'smart resource management' to the notion of relationships, which conjures up images of ecological context. Nearly all sustainable-agriculture research and practice is devoted to the human becoming smarter than in the past. What we are really after is to draw on the 'intelligence' embedded in the ecological context, which goes far beyond whatever 'smart' capabilities the human can muster.

We have said around here that we are working on perennial polyculture for grain production. I have never been quite comfortable with the expression and usually lace the phrase with humour when saying it aloud. I have done this almost unconsciously, because the expression carries a certain brittleness, a certain 'smartness', and emphasises *our* clever agronomic arrangement which forces us to feel more informed than we deserve to be. It has never captured the spirit of what we are about. Saying that we are out to build a domestic prairie featuring grain production shifts the focus from us to nature as the reservoir of intelligence. Attention is then drawn away from us as major agents of manipulation and we acknowledge that we are students of nature who are bent on discovery, students who want to build arrangements that imitate the current local results of the long evolutionary process.

The two questions—'What was here?' and 'What will nature require of us?' still dictate the terms of The Land Institute's research agenda. As we receive answers to these two questions, we gain insight into a third implied question articulated by Wendell Berry, 'What will nature help us do here?' I doubt that the agricultural researchers or farmers, maybe all the way back to Plato and Job, asked such questions. If all agricultural species, from corn plants to Holstein cows, are regarded as the property of humans rather than relatives of wild things that had most of their evolution in a context not of our making, then the best we can hope for is to become 'smart resource managers'. From my point of view, that's too limiting to prevent soil erosion, or, for that matter, any other desecration of the agricultural landscape or, ultimately, of us humans.

We can cheer the progress being made by farmers who are adopting practices that are less harmful to the soil and water. Yet we know that for agriculture to be *sustainable*, we must make a mental shift in how we see the world. Sir Albert Howard of England said we should farm like the forest. He might just as well have said we should farm like the prairie. Our work to develop several domestic mimics of the prairie mosaic signals our willingness to step out of the 'smart resource management' paradigm into a world into which we set out to discover the best fit with nature, both agriculturally and economically.

The economic order which I hope will emerge from these values will be the result of our becoming better community accountants. I believe our future will depend on a shift to accounting as a most important and interesting discipline. An accountant is a student of boundaries; in a small way modern accountants are forced to think about both the nature of boundaries and the classification of boundaries. From Maine to California, ecological accountants will have to think about what we should allow through the boundaries of various communities and at what rate. An input and output analysis forces us to examine the dynamics within those boundaries if we are to prevent squander and pollution and use efficiently the materials and energy within.

The elegant solutions John Todd talks about will be derivatives of these values and different ways of thinking about *efficiency* and *sufficiency*. No major programme, no public policy or project can account for all of the ecosystems from the cold deserts to the tropical rainforests and the Nebraska prairie in between. These evolving human communities, loaded with graduates who majored in Homecoming, perhaps will have to be engaged in a constant struggle of *quiet* succession from the power structure which has destroyed natural ecosystems, community, and communion during the extractive process. What the power structure knows, at some level, is that the reward for destroyed communion is more power. Community economics is the economics of place where 'elegant solutions can be predicated on the uniqueness of place'.

Chapter Twelve

Indigenous Rights—Colombia's Policy for the Amazon

Peter Bunyard

IT IS IMPRESSIVE TO TRAVEL the 500 kilometres upstream from La Pedrera, close to the border with Brazil, to the old penal colony at Araracuara, plum in the middle of the Colombian Amazon. Except for the small clearings of the occasional Indian settlement, the forest on either bank of the 1km-wide Caquetá river is intact. Moreover, flying over the region shows a continuous expanse of forest stretching as far as the eye can see. For once, official data appear to be correct in stating that of the 38.5 million hectares total area of the Colombian Amazon, more than three-quarters is still forested.

In February 1989 I went to Colombia at the invitation of Dr Martin von Hildebrand, head of Indigenous Affairs in the Colombian Government, to see for myself the dramatic changes taking place in the Amazon as a result of the Government's policy to grant title to the lands of its indigenous inhabitants. One of my first trips was out of Mocoa, in the Upper Putumayo, from where I walked along a winding, precipitous mule track to visit an Indian community living on a small plateau overlooking the Caquetá river as it plunged down the foothills of the Andes to the Amazon plains below. There, I saw the effects of centuries of exposure to missionaries and more recently of colonists upon the 400-strong community of quechua-speaking Inga indians. That was to be a reminder of how destructive our western civilization could be not only of cultures but also of the environment.

The Colombian Amazon has had its fair share of missionaries in recent years and of exploitation by various dealers, many from Brazil and Peru. Yet, with its vast expanse, it has managed to avoid much of the environmental degradation that has characterized great stretches of the Andean foothills, whether on the Amazon side or moving down westwards to the Pacific Ocean. Indeed, many of the Indian communities, especially those of the lower reaches of the Caquetá river, have retained much of their traditional cultures and the question has to be asked whether this retention is due to the environment around being intact, or whether the environment is intact because the Indians still live within a traditional culture.

By the time I arrived, the Indians of the Colombian Amazon had received title from the Government to 12 million hectares, the bulk of the land being in the central region of Colombia's Amazon territory. This remarkable act was followed half a year later by the granting of title to a further 6 million hectares, thus putting into the hands of 70,000 Indians a contiguous area of land that was more than three-quarters the size of Great Britain and was the largest government-granted territory to indigenous peoples of any country.

Colombia had, therefore, led the world in recognizing its obligations to its indigenous peoples. Moreover, it had done so in full recognition of its intention that its Amazon rainforest should be preserved, at least the bulk of what was left. Colombia had instituted the granting of title to the Indians of the Amazon without any pressure from outside. In fact, the Colombian Government appears to have fulfilled certain of the requests of the South American Indian Organization – COICA – that conservation of the Amazon rainforest should be directly linked to recognition of indigenous rights to land.

During my stay in the Amazon I joined up with Indian leaders who were attending a special meeting with government officials and the Puerto Rastrojo Foundation to discuss how the new policies for the Amazon would affect their communities. During the week of intense discussions it became clear that the Indians accepted fully the role put upon them by Government for caring for the rainforest. In return, the Government had enshrined in law the fundamental rights of the Indians to their own cultures and traditions. In fact, these were to be encouraged, being considered part and parcel of ecosystem conservation.

Colonization, both close to the foothills of the Andes, as they lead to the extensive plains of the Amazon Basin, as well as along the northern perimeter where forest gives way to savannah, has invariably destroyed the forest. According to Government statistics the amount of forest lost over the past 40 years amounts to nearly 6 million hectares, hence an average total of over 100,000 hectares a year.[1]

However, the great bulk of the colonization took place during the 1940s and 50s as a result of the political violence which resulted in a million casualties and led to peasants fleeing their homes. For the time being, the pace of new colonization has stagnated, although hardly stabilized, and the attrition of forest could continue unless sound schemes are elaborated for holding still the colonization front. Today, the colonist population of Colombia's Amazon region amounts to nearly half a million as compared to the 70,000 Indians from some 50 different ethnic groups.[2]

Land distribution in the Colombian Amazon is skewed, with 25 million hectares in the hands of the indigenous communities and 6 million in the hands of the colonists with their 7 times greater population. Some 8 million hectares of Colombia's Amazon are barely inhabited at all, the remaining forest now forming part of Colombia's forest reserves.

The Colombian Government's far-reaching policy to protect the remaining rainforest from further destruction has been carried out despite intense pressures to develop its Amazon, from agricultural, mining and logging interests. What makes the Colombian initiative unique is that the land put officially into Indian hands is based on their extensive use of the forest and not on some western notion of the land being subdivided into small agroforestry plots.

Over the past decade, lawyers, anthropologists, biologists, politicians and Indian leaders have worked together to find a legal formula that would enable the Government to grant all the Indian groups of the Amazon inalienable rights to their traditional lands. Through resurrecting old colonial laws, Colombia has now found a legitimate mechanism that concedes the entitlement of indigenous communities to land to which they had *a priori* rights and equally grants them considerable autonomous status within Colombian sovereignty. For instance, indigenous communities in Colombia now have full rights to bring up their children according to their own traditions and in their own languages, to practise their own traditional medicines, and to pursue their own traditional systems of authority in holding their communities together.

During the centuries of colonial rule, the Spanish Government conceded certain areas to the indigenous population as 'resguardos'. The law pertaining to the resguardo holds that the indigenous peoples residing within and utilizing the area of the resguardo own communal land rights which are inalienable. In modern terms, the state cannot capriciously bring about the dissolution of the resguardo nor parcel it out without coming into direct conflict with the laws of the Constitution. Equally, since the land is held communally, individual members of the community cannot acquire or sell off any part of the resguardo. Any such transfers would require the agreement of more than three-quarters of the adults of the community. A reservation, on the other hand, does not provide the same guarantees insofar as the state can bring about its dissolution without contravening state laws.

Since 1961, Colombia has instituted 221 resguardos and 19 reservations for its indigenous peoples. Some 158,000 Indians now hold land rights over the country to nearly 26 million hectares. An equal number of Colombian Indians, from a wide variety of ethnic groups, inhabit resguardos granted during Spanish colonial times, prior to 1821. The total land area of these ancient resguardos, none of which are in the Amazon, amounts to no more than 400,000 hectares.

Colombia has an estimated 450,000 indigenous peoples, approximately double the numbers officially recognized as being indigenous in neighbouring Brazil. Of that total, some 85,000 have yet to receive title to lands. On the other hand, although the Government recognizes that 210,000 Indians in the Andean region of the country have title to lands that go back to colonial times, much of the land is occupied by non-Indian landowners who have settled there over the past two centuries. The Government is buying back some of this land for the Indians, but the process is slow owing to the high cost involved and the vexed problem of persuading long-established landowners to relinquish what they by now consider to be their own lands.

Why should the Colombian Government have taken the unusual step of stating that the rainforest is best looked after by its indigenous peoples? One of the main reasons is undoubtedly the manifest destruction of the rainforest at the hands of the colonists and other exploiters since World War II. Of its total 1.147 million square kilometres, some 0.78 million square kilometres have soils and climate that best support forest. However, today, only 0.531 million square kilometres are forested, some 37.7 million hectares having been cut down between 1960 and 1984. Deforestation over the past 30 years has oscillated between 660,000 and 880,000 hectares a year. Such a rate, if continued, would see most of Colombia's rainforest vanish within 50 years.

The Colombian Amazon, meanwhile, has been losing forest at the rate of 100,000 hectares per year, a rate which, if continued, would deforest the entire region in several hundred years. However, any complacency would be wholly misplaced, since once the forests have gone in other parts of the country, the Amazon could well become the focus of attention for foresters and colonists alike.

In Colombia, as in other Latin American countries, most of the forest is destroyed to make way for cattle ranching. In 1987, the Geographic Institute of Augustin Codazzi (IGAC) carried out a study into actual land use and its potential in Colombia. It found that 12.7 per cent of Colombian territory was well-suited to agriculture but that only 4.6 per cent was so used. The situation was reversed for cattle ranching with more than one-third of the country dedicated for such a purpose when at most it should have been carried out on just 16 per cent. Where tropical forest had gone, the land in 93 per cent of the cases had been turned over to cattle ranching. The carbon dioxide emitted each year from the destruction of Colombia's forests puts per capita emissions on a par with an industrialized nation such as Britain.

How much is it chance that the Indian communities of the Amazon, like their Embera counterparts in the Choco, have lived with the forest ecosystem without destroying it? The answer lies in their deep understanding of the reciprocal, give-and-take relationship between the forest and the species it harbours. The Indians are not ecologists by having studied science, but through their culture. That understanding prevents traditional communities from committing excesses against the forest and trying to extract more than it can give. As studies of the Yucuna Indians around the Miritiparana river of the eastern part of the Colombian Amazon have shown, the Indians basically use the forest in an extensive fashion.

Tomas Walschburger and Patricio von Hildebrand of the Puerto Rastrojo Foundation have found that the Yucunas typically use at least 16 types of ecosystem to furnish them with their basic necessities.[3] In the orchard, covering some 2 to 3 hectares around the communal maloca, they cultivate fruit trees such as the chontaduro palm, avocado, papaya, lemon, mango, maraca (related to cacao) and others. Meanwhile the nearby forest provides materials from some 168 different plant species: wood and vines for building the maloca; wood for canoes; and medicinal plants as well as plants for ritual and shamanistic purposes and for preparing curare. The same area also provides game, especially small animals and birds. Further from home, the Yucuna families go to the distant forest, encompassing many thousands of hectares, to hunt and to collect materials not available in the nearby forest. The nearby lake provides fish during the summer months when the waters are low, while the cananguchales, those areas that are permanently or seasonally flooded, are important for the canangucho tree (*Mauritia flexuosa*) which has a nutritious fruit and attracts game.[4]

The chagras or garden areas form a relatively small part of the territory the Indians need to subsist. Since they can be used for two or three seasons at most before being abandoned, the area around the maloca is dotted around with chagras at different stages of regeneration. The use of fire to burn off the wood and constant weeding out of saplings during the first few months of making a clearing leads to the regeneration to mature forest taking more than 100 years, compared to the 40 years of regeneration following a natural clearing.

Other areas of the ecosystem used by the Indians include flood plains which provide fish as well as palm leaves for thatching, rivers for fish and for the eggs of turtles which lay in sandy beaches, and, finally, natural salt licks which at certain seasons attract large game such as tapirs and deer.

The Indians are very selective in choosing the areas for making their gardens, the kind of forest that is growing providing a good indicator of soil fertility and suitability. With rare exception they make their clearings in primary forest, which must have little understorey and relatively few surface roots, since well-developed root mats hinder clearing and indicate a nutrient-deficient soil. The soil should be between sand and clay: the sand is good for yucca and other tuberous plants, while clay favours plantains, coca and fruit bushes. Unlike the colonists, the Indians avoid the vegas, or flood plains along the river banks which, although more fertile, can be washed away.

To some extent, the creation of the chagra mimics the natural processes of decay and regeneration. Natural clearings appear when a few trees fall, leaving open spaces of up to 500 square metres. Many trees in the lower storey survive the fall of the larger trees and protect the soil from direct sunshine while letting enough light through to stimulate the rapid germination of pioneer species that lie dormant in the soil. Some fallen trunks sprout new shoots and trees eventually grow back. A natural clearing also harbours many animals which deposit the seeds of woody species in their excrement.

In contrast to the natural clearing, when the Indians create their chagras they fell all the trees and the sudden insurgence of direct sunlight kills young plants and stunts the growth of others. Young saplings of the understorey and shoots that spring from fallen trunks die in the bright sunlight. About three months after cutting down the trees, the Indians fire the fallen wood, killing the shoots of many pioneer species, such as cecropia. However, the chagra provides the ideal environment for herbaceous plants: hence, while the seeds of pioneer species predominate when the garden is first cleared, after two years of selecting out more than 90 per cent of the seeds belong to herbaceous species.[5]

In making their chagras, the Indians are well aware of the importance of the intact forest around as the regenerator or mother of fertility. They purposely do not burn or cultivate the boundaries of their clearings: furthermore they grow fruit trees as well as subsistence crops such as yucca, peppers, tobacco and coca. The fruit attracts birds and bats whose droppings contain a variety of seeds, which, in germinating, speed up the process of regeneration. Recent studies of the movements of birds and bats validates the Indians' practice of keeping the size of chagras to approximately half a hectare, since if the clearings are too big, the animals will not venture into them. Rather than create one large chagra to provide more food, the Indians thus create one or more new ones.

The Indians harvest crops such as yucca for two years. Then, as the soil loses its fertility and weeds begin to encroach on the clearing, they move on, growing crops in a new chagra, but returning to the old one to collect fruit and hunt game. An abandoned chagra may provide fruit for up to 15 years. Initially, the trees grow very quickly in an abandoned chagra. Each hectare produces about 15 tonnes of biomass for the first couple of years. After 60 years, the total biomass, taking into account the natural attrition of trees, is some 150 tonnes per hectare. In contrast, clearings made by bulldozing away all vegetation show virtually no regeneration and 10 years later may still have no more than 2 tonnes per hectare of biomass, mostly made up of herbaceous species.

For its value to human beings, the forest cannot be judged in terms of the timber it might provide or the capability of its soil to sustain agriculture. It is easy, too, to overestimate the quantities of forest products that can be extracted. Indeed, estimates of the amount of land required by traditional Indians for all their activities, including hunting, fishing, gathering and gardening, are found to lie in the region of at least 1000 hectares for every man, woman and child.[6]

The Indians do not see the forest as being there solely for their benefit or for the benefit of human beings in general. On the contrary, embedded in their view of the world is the notion that each living part of the rainforest must be given the opportunity to exist in order to sustain the integrity of the whole and without that wholeness will come disease, disaster and death.

Martin von Hildebrand found from his work with the Tanimuca Indians of the Upper Piraparana that they view nature and the whole of creation as participating in an intricate network of exchanges and reciprocity. Their interpretation is that the forms seen in the forest are the external manifestation of an entity which can be described as *thought* and which anthropologists tend to translate as *essence* or *energy*. According to the Indians, the amount of *thought* is limited and, therefore, it has to be recycled among the different species, each having its right quota.

The *thought* emanates chiefly from the Sun, or from a place in the east where the Sun rises and everything in the world originated. Each group of animals, plants and people needs a certain amount of *thought* to survive. For this, they have guardians whose role it is to see that each group has enough *thought* and that nobody steals more than their share. When people, or indeed other animals, consume plants or animals they are taking *thought* from these groups and accumulating it in their own bodies. The guardian of all hunted animals is represented by the *anteater*, the guardian of wild fruit by the *tapir*, the guardian of crops by the *anaconda* and the *cicada*. The guardian of the jungle is the *jaguar*, whereas that of humans is the *jaguar-man* or shaman.

When people or other living creatures become sick and die they release *thought* which then can recycle and be trapped by animals, plants or, indeed, people. When people hunt or collect plants they must do so under the direction of their guardian or shaman so as to obtain *thought* or *energy* for their group. Meanwhile, the guardians of the animals or plants hunt people in return by sending sickness and illness or by causing accidents. Hence, if a person consumes too much of a certain plant or animal his *thought* will become visible to the guardians and appear as what he has consumed. Thus, although he will still see himself as a human being, inside he will have become more like an animal or plant, depending on what he has consumed and how much. The guardians will then hunt him down. To avoid such dangers the Indian must be careful what he eats and how much; at the same time he must release *thought* through rituals and fasting.

The shaman is a person who has been trained to understand the environment; he is an ecologist by culture. He establishes contact with the guardians of the plants and animals either by entering into altered states of awareness induced by hallucinogenic drugs or by fasting. He then negotiates with them which animals or plants his community can use and offers in exchange coca powder, or the *thought* of those who have died, be it of his own or of other communities. On the basis of his negotiations he tells his people where and what they can hunt as well as how much.

The permission varies with the seasons, with the animals, their reproductive cycles and the use they make of different areas of the forest. Every two or three months the community celebrates communal rituals to collect the extra *thought* that has accumulated in the group so that it can be returned to the respective guardians. After the ritual the shaman imposes food and sexual restrictions. These restrictions follow other rituals, such as ones concerning healing, birth, death and especially initiation rites. Taken together, these activities lead to effective control over the communal demand for natural resources.

The dynamic of the forest and the interchange of matter between one species and another, including that of the life-force, provides the Indians, with a ready model of their own existence within the community of the maloca. Hence, the local economy, both within the community and between neighbouring communities, relies heavily on the principle of exchange and reciprocity, both among themselves and with the rest of nature.

Contrary to the system engendered by the market economy, in which a person's status increases with his wealth and possessions, the traditional economy of the Indians is *anti-market.* Indeed, the person who accumulates is evidently one who lacks social relations with others and has no one with whom to share. Surpluses, as and when they arise, serve to establish relations informally within the community and more formally with neighbouring communities invited in during rituals and festive occasions, or when there is heavy work to be done, such as the making of new chagras in the forest or in constructing a new maloca. Marriage alliances between communities also play an integral part in the processes of exchange and reciprocity.

The Indians believe that both the animals and plants are similar to them, maintaining among each other alliances involving exchange and reciprocity. They thus respect the territories and way of life of their fellow creatures; they know how each behaves and they fear the consequences of any abuse. When living within their traditions, therefore, the Indians will hunt or exploit other beings only after asking for permission under the guidance of the shaman, who is constantly evaluating the state of the environment.

Authority within the traditional community resides with the shaman, with the man responsible for the smooth-running of the communal house and, to a lesser degree, with the cantor whose role it is to run the rituals and dances. In general, these three individuals are brothers and their authority is inherited.

The coming into contact with white dealers and the savage exploitation by the rubber barons, which led to the decimation of many indigenous communities in the Colombian Amazon at the turn of the century, the establishment of Catholic and Protestant missions, and the need for goods, such as steel axes, machetes, outboard motors and the fuel to run them, have all contributed to the eroding of traditions and customs and a loss of respect for the traditional leaders. The priests tried to get the Indians to give up their communal living and forced parents to send their children into the mission schools where the process of acculturation could be finally accomplished. The Indians, in fact, became accomplices in the decimation of wildlife during the 1950s and 1960s, when the trade in furs was at its height. Many of them have also been involved in the overfishing of the rivers, selling the fish to white dealers, who have established refrigeration units at various strategic points, so as to supply Bogotá.

The issue of converting parts of the Amazon into vast coca plantations with processing laboratories hidden away in the jungle has also raised its ugly head. During the late 1970s, some Indian communities of the Colombian Amazon did become embroiled in producing coca for the market—especially for the production of 'basuco' or 'crack'. That phase, like the hunting of skins, has somewhat passed, particularly as production has increased in Bolivia and Peru. The Indian leaders believe that, with the proper recognition of their rights and the re-affirmation of their traditions, the temptation to earn from growing coca for the market will diminish even more. Indeed, the production of coca is, as commentators such as COICA have pointed out, primarily a socio-economic issue with peasants in particular struggling desperately to earn sufficient to survive.

The destruction of the forest, some 250,000 hectares in Peru alone, for coca production—as in the Alto Huallaga valley—is directly linked to consumption in the United States and increasingly in Europe. The reaction of the United States to attack the source of the drug rather than its own insoluble social problems may be understandable, but its efficacy remains very much in question.

In fact, US scorched-earth policies, as practised during the 1980s in the marijuana-growing areas on the slopes of the Sierra Nevada de Santa Marta in the north-east of Colombia, had a devastating effect. The spraying of gylphosphate (*Roundup*) from helicopters, regardless of its effect on human beings, inevitably led to increased destruction of the forest, since those who lost their crop immediately attacked another area in their desperate attempt for economic survival. Within a matter of 20 years, more than half the rich rainforests of the Sierra Nevada had gone up in smoke. Colombians point out the irony of the situation inasmuch as the marijuana consumers of the United States now obtain most of their supplies from California.

Gradually, the traditional leaders are regaining their authority, an authority that is now recognized by the state itself. Indeed, in what must be unique in the world, the Colombian Government is not only actively encouraging the indigenous communities of the Amazon to return to their traditions and cultures, but has given them the space and authority to do so. Moreover, it expects that the return to a traditional way of life will lead to the protection of the rainforest. At least in the Amazon, Colombia appears to have got its values straight—and the word is spreading. Bolivia is now seeking advice from Colombian lawyers on how to create resguardos for the Indians of its Amazon region. Who else will follow suit?

Chapter Thirteen

Recovering Diversity—A Future for India

Vandana Shiva

Diversity and Sustainability

The two principles on which the production and maintenance of life are based are: (*a*) the principle of diversity and (*b*) the principle of symbiosis and reciprocity, often also called the law of return.

The two principles are not independent but interrelated. Diversity gives rise to the ecological space for give and take, for mutuality and reciprocity. Destruction of diversity is linked to the creation of monocultures and, with the creation of monocultures, the self-regulation and decentralized organization of diverse systems gives way to external inputs and external and centralized control. Schematically, the transformation can be illustrated as below:

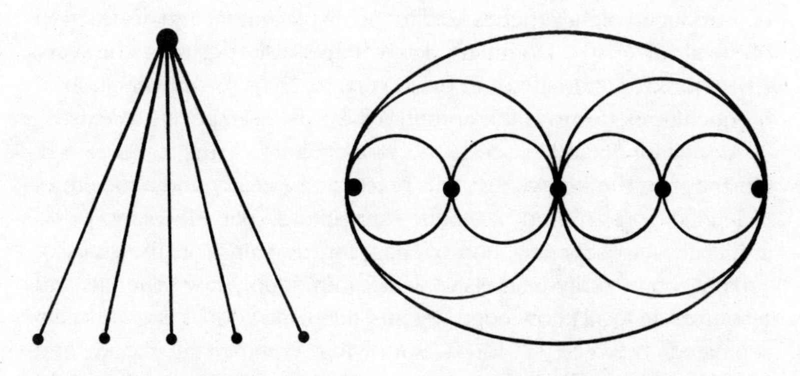

Monocultures are associated with external inputs, centralized regulation and high vulnerability to ecological breakdown

Systems based on diversity are associated with decentralized self-regulation and high resilience

Sustainability and diversity are ecologically linked, because diversity offers the multiplicity of interactions which can heal ecological disturbance to any part of the system. Non-sustainability and uniformity mean that a disturbance to one part is translated into a disturbance to all other parts. Instead of being contained, ecological destabilization tends to be amplified. Closely linked to the issue of diversity and uniformity, is the issue of productivity. Higher yields and higher production have been the main reason for the introduction of uniformity and the logic of the assembly line. The imperative of growth generates the imperative for monocultures. Yet this growth is, in large measure, a socially constructed, value-laden concept. It exists as a 'fact' by excluding and erasing the facts of diversity and production through diversity.

The disappearance of diversity

Over and above rendering local knowledge invisible by declaring it non-existent or illegitimate, the dominant system also makes alternatives disappear by erasing and destroying the reality which they attempt to represent. The fragmented linearity of the dominant knowledge disrupts the integrations between systems. Local knowledge slips through the cracks of fragmentation. It is eclipsed along with the world to which it relates. Dominant scientific knowledge thus breeds a monoculture of the mind by making the space for local alternatives disappear, very much as monocultures of introduced plant varieties lead to the displacement and destruction of local diversity. Dominant knowledge also destroys the very *conditions* for alternatives to exist, very much as the introduction of monocultures destroys the conditions for diverse species to exist.

As metaphor and process, the monoculture of the mind is best illustrated in the knowledge and practice of forestry and agriculture. 'Scientific' forestry and 'scientific' agriculture split the plant world artificially into separate, non-overlapping domains, on the basis of separate commodity markets to which they supply raw materials and resources. In local knowledge systems, the plant world is not artificially separated between a forest supplying commercial wood and agricultural land supplying food commodities. The forest and the field are an ecological continuum and activities in the forest contribute to the food needs of the local community, while agriculture itself is modelled on the ecology of the tropical forest. Some forest dwellers gather food directly from the forest, others do swidden agriculture* within the forest, while many communities practise agriculture outside the forest, but depend on the fertility of the forest for the fertility of agricultural land.

In the 'scientific' system, which splits forestry from agriculture and reduces forestry to timber and wood supply, food is no longer a commodity related to forestry. The cognitive space that relates forestry to food production, either directly or through fertility links, is therefore erased with the split. Knowledge systems which have emerged from the food-giving capacities of the forest are, therefore, eclipsed and finally destroyed, through neglect and aggression.

Most local knowledge systems have been based on the life-support capacities of tropical forests, not on their commercial timber value. These systems lie in the blind spot of a forestry perspective that is based exclusively on the commercial exploitation of forests. If some of the local uses can be commercialized they are given the status of 'minor products', with timber and wood being treated as the 'major product' in forestry. The creation of fragmented categories thus blinkers out entire spaces in which local knowledge exists, knowledge which is far closer to the life of the forest and more representative of its integrity and diversity. Dominant forestry science has no place for the knowledge of the Hanunoo in the Philippines, who divide plants into 1,600 categories, of which trained botanists can distinguish only 1,200.[1] The knowledge base of the cropping systems of the Lua tribe in Thailand, based on 116 crops, is not counted as knowledge, either by dominant forestry, which sees only commercial wood, or dominant agriculture, which sees only chemically intensive agriculture. Food systems based on the forest, either directly or indirectly are, therefore, non-existent in the field of vision of a reductionist forestry and a reductionist agriculture, even though they have been and still are the sustenance base for many communities in the world. For example, the rainforests of South-east Asia supply all the food needs of the Kayan, the Kenyah, the Punan

*Swidden agriculture is also called 'shifting cultivation' and refers to rotational cropping in a forest ecosystem, with the location being shifted to allow re-growth of the forest for regenerating soil fertility.

Bah, Tanyjong, and the Penan, who gather food from the forest and practise swidden agriculture. The Tiruray people depend on the wild flora of the forests as a major source of food and other necessities.[2] Plant supplies are gathered mostly from the surrounding forest and some 223 basic plant types are regularly exploited. The most important food items are mushrooms (kulat), ferns (paku) and the hearts of various plants (ubot) which include bamboo shoots, wild palms, and wild bananas; 25 different varieties of fungi are eaten by the Kenyah and 43 varieties are eaten by the Iban.[3] Sago, the staple of the Penan of Borneo, is the starch obtained from the pith of a palm tree called *Eugeissone utilis.* On New Guinea as a whole (Irian Jaya and Papua New Guinea together), 100,000 sago eaters produce 115,000 metric tons of sago each year.[4]

Ethnobotanical work among India's many diverse tribes is also uncovering their deep, systematic knowledge of forests. The diversity of forest foods used in India emerges from this knowledge. In southern India, a study conducted among the Soliga in the Belirangan hills of Karnataka shows that they use 27 different varieties of leafy vegetables at different times of the year and a variety of tubers, leaves, fruits and roots are used for their medicinal properties. A young illiterate Irula boy from a settlement near Kotagiri identified 37 different varieties of plants, and gave their Irula names and their different uses.[5]

In Madhya Pradesh, although rice (*Oryza sativa*) and lesser millets (*Panicum miliaceum*, *Eleusine coracana* and *Paspalum scrobiculatum*) form the staple diet of the tribals, almost all of them supplement it with seeds, grains, roots, rhizomes, leaves and fruits of numerous wild plants which abound in the forests. Grigson noted that famine has never been a problem in Bastar, as the tribes have always been able to draw half of their food from the innumerable edible forest products. Tiwari[6] prepared a detailed list of wild plant species eaten by the tribals in Madhya Pradesh. He has listed 165 trees, shrubs and climbers. Of these, the first category contains a list of 31 plants whose seeds are roasted and eaten. There are 19 plants whose roots and tubers are eaten after baking, boiling or processing; 17 whose juice is taken fresh or after fermenting; 25 whose leaves are eaten as vegetables and 10 whose petals are cooked as vegetables. There are 63 plants whose fruits are eaten raw, ripe, or roasted or pickled and there are five species of *Ficus* which provide figs for the forest-dwellers. The fruits of the thorny shrub *Pithcellobium dulce* (*Inga dulcis*), also called jungle jalebi, are favourites with the tribals. The sepals of mobwa are greedily eaten and also fermented for liquor. *Morus alba*, the mulberry, provides fruit for both people and birds. The berries of *Zizyphus mauritiana* and *Z. oenoplia* provide delicious fruit, and have been eaten by jungle dwellers from the Mesolithic period onwards.[7]

In non-tribal areas, too, forests provide food and livelihood through critical inputs to agriculture, through soil and water conservation, and through inputs of fodder and organic fertilizer. Indigenous silvicultural practices are based on sustainable and renewable maximization of all the diverse forms and functions of forests and trees. This common silvicultural knowledge is passed on from generation to generation, through participation in the processes of forest renewal and of drawing sustenance from the forest ecosystem.

In countries like India, the forest has been the source of fertility and the renewal of agriculture. The forest as a source of fodder and fertilizer has been a significant part of the agricultural ecosystem. In the Himalayas, the oak forests have been central to the sustainability of agriculture. In the Western Ghats, the 'betta' lands have been central to the sustainability of the ancient spice gardens of pepper, cardamon, and areca nut. Estimates show that over 50 per cent of the total fodder supply for peasant communities in the Himalayas comes from forest sources, with forest trees supplying 20 per cent.[8] In Dehradun, 57 per cent of the annual fodder supply comes from the forests.[9] Besides fodder inputs, forests also make an important contribution to hill farming in the use of plant biomass as bedding for animals. Forests are the principal source of fallen dry leaf-litter and lopped green foliage of trees and herbaceous species, which are used for animal bedding and composting. Forest biomass, when mixed with animal dung, forms the principal source of soil nutrients for hill agriculture. According to one estimate, 2.3 metric tonnes of litter and manure are used per hectare of cultivated land annually.[10] As this input declines, agricultural yields also go down.

The diverse knowledge systems which have evolved with the diverse uses of the forest for food and agriculture were eclipsed with the introduction of 'scientific' forestry, which treated the forest only as a source of industrial and commercial timber. The linkages between forests and agriculture were broken and the function of the forest as a source of food was no longer perceived.

The destruction of biological diversity is intrinsic to the very manner in which the reductionist forestry paradigm conceives of the forest. The forest is *defined* as 'normal' according to the objective of managing the forest for maximizing production of marketable timber. Since the natural tropical forest is characterized by richness in diversity, including the diversity of non-marketable, non-industrial species, the 'scientific forestry' paradigm declares the natural forest to be 'abnormal'. In Schlick's words, forest management implies that 'the abnormal conditions are to be removed'[11] and, according to Troup: 'The attainment of the normal forest from the abnormal condition of our existing natural forest, involves a certain temporary sacrifice. Generally speaking, the more rapid the change to the normal state, the greater the sacrifice; for example, the normal forest can be attained in one rotation by a series of clear fellings with artificial regeneration, but in an irregular, uneven-aged forest this means the sacrifice of much young growth which may be unsaleable. The question of minimizing the sacrifice involved in introducing order out of chaos is likely to exercise our minds considerably in connection with forest management.'[12]

The natural forest, in its diversity, is thus seen as 'chaos'. The man-made forest is 'order'. 'Scientific' management of forests, therefore, has a clear anti-nature bias, and a bias towards industrial and commercial objectives, for which the natural forest must be sacrificed. Diversity thus gives way to uniformity of even-aged, single-species stands, and this uniformity is the ideal of the normal forest towards which all silvicultural systems aim. The destruction and dispensibility of diversity is intrinsic to forest management guided by the objective of maximizing commercial wood production, which sees non-commercial parts and relationships of a forest ecosystem as valueless—as weeds to be destroyed. Nature's wealth, characterized by diversity, is destroyed to create commercial wealth, characterized by uniformity. In biological terms, tropical forests are the most productive biological systems on our planet.

A large biomass is generally characteristic of tropical forests. The quantities of wood, especially, are large in tropical forests and average

about 300 tons per hectare compared with about 150 tons per hectare for temperate forests. However, in the reductionist commercial forestry, the overall productivity is not important, nor are the functions of tropical forests in the survival of tropical peoples. It looks only for the industrially useful species that can be profitably marketed and measures productivity in terms of industrial and commercial biomass alone. It sees the rest as waste and weeds. As Bethel, an international forestry consultant, states, referring to the large biomass typical of the forests of the humid tropics:

'It must be said that from a standpoint of industrial material supply, this is relatively unimportant. The important question is how much of this biomass represents trees and parts of *preferred species that can be profitably marketed*... By today's utilisation standards, *most of the trees, in these humid tropical forests are, from an industrial materials standpoint, clearly weeds.*'[13] [my italics]

The industrial materials standpoint is the standpoint of capitalist, reductionist forestry, which splits the living diversity and democracy of the forest into commercially valuable dead wood and destroys the rest as 'weeds' and waste. This 'waste', however, is the wealth of biomass that maintains nature's water and nutrient cycles and satisfies needs for food, fuel, fodder, fertilizer, fibre, and medicine in agricultural communities.

Just as the 'scientific' forestry paradigm excludes the food-producing functions of the forest and destroys the forest diversity as 'weeds', 'scientific' agriculture, too, destroys species which are useful as food, even though they may not be useful on the market.

The green revolution has displaced not just seed varieties but entire crops in the Third World. Just as peoples were, seeds were declared 'primitive' and 'inferior' by the green revolution ideology, food crops were declared 'marginal', 'inferior' and 'coarse-grained'. Only a biased agricultural science rooted in capitalist patriarchy could declare nutritious crops like ragi and jowar 'inferior'. Peasant women know the nutritional needs of their families and the nutritive content of the crops they grow. Among food crops, they prefer those with maximum nutritional value to those with a value in the market. What have usually been called 'marginal crops' or 'coarse grains' are nature's most productive crops in terms of nutrition. That is why women in Garhwal continue to cultivate mandua and women in Karnataka cultivate ragi in spite of all attempts by state policy to shift to cash crops and commercial foodgrains, to which all financial incentives of agricultural 'development' are tied. Table 1 illustrates that what the green revolution has declared 'inferior' grains are actually superior in nutritive content to the so-called 'superior' grains: rice and wheat. A woman in a Himalayan village once told me, 'Without our mandua and jhangora, we could not labour as we do. These grains are our source of health and strength.'

Table 1
Nutritional content of different foodcrops

	Protein (gms)	Minerals (100 gms)	Ca (mg)	Fe (100 gms)
Bajra	11.6	2.3	42	5.0
Ragi	7.3	2.7	344	6.4
Jowar	10.4	1.6	25	5.8
Wheat (milled)	11.8	0.6	23	2.5
Rice (milled)	6.8	0.6	10	3.1

Not being commercially useful, people's crops are treated as 'weeds' and destroyed with poisons. The most extreme example of this destruction is that of bathua, an important green leafy vegetable, with a very high nutritive value and rich in vitamin A, which grows as an associate of wheat. However, with intensive chemical fertilizer use bathua becomes a major competitor of wheat and has been declared a 'weed' that is killed with herbicides. Forty thousand children in India go blind each year for lack of vitamin A, and herbicides contribute to this tragedy by destroying the freely available sources of vitamin A. Thousands of rural women who make their living by basket and mat-making with wild reeds and grasses are also losing their livelihoods, because the increased use of herbicide is killing the reeds and grasses. The introduction of herbicide-resistant crops will increase herbicide use and thus increase the damage to economically and ecologically useful plant species. Herbicide resistance also excludes the possibility of rotational and mixed cropping, which are essential for a sustainable and ecologically balanced agriculture, since the other crops would be destroyed by the herbicide. US estimates now show a loss of US$ 4 billion per annum due to crop loss as a result of herbicide spraying. The destruction in India will be far greater, because of higher plant diversity and the prevalence of diverse occupations based on plants and biomass.

The one-dimensional perspective of dominant knowledge is rooted in the intimate links between modern science and the market. As multidimensional integrations between agriculture and forestry at the local level are broken, new integrations between non-local markets and local resources are established. Since economic power is concentrated in these remote centres of exploitation, knowledge develops according to the linear logic of maximizing the flow of cash and commodities and becomes blind to the cyclical flows at the local level. The integrated forest and farm gives way to the separate spheres of forestry and agriculture. The diverse forest and agricultural ecosystems are reduced to 'preferred' species by the selective annihilation of species diversity, which is not 'useful' from the market perspective. Finally, the 'preferred' species themselves have to be engineered and introduced on the basis of 'preferred' traits. The natural, native diversity is displaced by introduced monocultures of tree and crops.

Towards an organic future

The destruction of diversity that has been the invisible aspect of growth and development has had high social and ecological costs. Ecologically, new instabilities and vulnerabilities have been introduced in the form of desertification, land-degradation, disease and pest outbreak, and chemical contamination of land and water. Socially, the diversity that has been lost has often been the basis of sustenance for local communities. Environmental degradation and the creation of poverty have thus gone hand in hand.

There are two options in India for correcting the development blunders of the past 20 years, which have destroyed the ecological capital that had been maintained over centuries. The first option is the technological fix that transnational corporations and national Governments offer. In this case, the biotechnology and genetic engineering solutions are being offered as 'organic'. Biotechnology has benefited from its inclusion under the 'biological' category, which has connotations of being ecologically safe. Indeed, the biotech industry has described its agricultural innovations as 'Ecology Plus'.

However, the biotech option does not offer an organic future. First, the new biotechnologies deepen the fragmentation and creation of uniformity that the green revolution, for example, produced. Secondly, the 'engineering' ethos of genetic engineering erodes the self-regulation and self-regeneration capacity of nature and local production systems. It creates further dependencies on external inputs and external management.

The foundations for an organic future are based on the permanent principles of diversity and reciprocity. The contributions to this future are not coming from corporate labs or government ministries. They are coming from the small initiatives of local communities.

Recovering diversity

At an altitude of 6,000 ft., deep in the Balganga Valley in Garhwal, lies Kangad, a hamlet of 200 families. In 1977, the already degraded forest of Kangad was marked for felling by the forest department. The women, who had to walk long distances for fuel, fodder and water, were determined to save the last patch of trees. The men of Kangad were employed by the forest department for felling operations. With the gender fragmentation of the interests of the village community—the women representing the conservation interests and the men representing the exploitation demand—launching the Chipko movement was not easy. The women contacted Bimla Bahuguna in Silyara, just 15 kms from Kangad. Bimla Behn, with Chipko activists Dhoom Singh Negi and Pratap Shihar, came to support the women's struggle. After four months of resistance, the women succeeded in saving their forest.

The women's organization, the Mahila Mandal Dal, then decided to regenerate the degraded forests. On the basis of the number of cattle owned by each family, contributions were raised to support a village forest guard who was paid Rs. 300 per month. For three years the arrangement worked and then failed because the watchman became inefficient and corrupt: he would allow some people to extract fodder and fuelwood. Once the women learnt of this, they unanimously decided to abolish the post of the forest guard and guard the forest themselves.

Now the Mahila Mandal has allocated duties to a group of village women. About 10 or 12 women are on duty every day, allocated in such a manner that the work is distributed among all the families. Thus, the duty for one family or group of women comes in a cycle of 15 to 20 days. As one woman said, 'On these days we leave our own work and protect the forest, because our oak trees are like our children.' Oak trees are now regenerating naturally in Kangad.

Once, when a Gujjar grazier allowed his goats to graze in the regenerated area, the women confiscated the goats and fined the Gujjar Rs. 200. Villagers are fined up to Rs. 50 per person for lopping the regenerating oak and Rs. 100 for cutting trees for firewood. On another occasion, when a fire threatened to destroy the forest, all the women joined hands to put out the forest fire. As one woman reported: 'The men were at home, but they decided to stay back rather than join with us to put out the fire. The men are least bothered about saving trees.' In 1986, the Mahila Mandal decided to assist the forest department in tree planting. They dug 15,000 pits, but found that the forest department wanted to plant only poplars. The women refused to plant this exotic, and forced the forest department to bring various indigenous fodder species instead.

The strength of nature and the strength of women are the basis of the recovery of the forest as commons in Kangad. The capital is not debt and aid. The market is not the guiding force. Nature's and women's energy are the capital, and local needs of water, food, fodder and fuel provide the organizing principle of managing a shared, living resource. This is merely a renewal of the conservation ethic and conservation work of hill women, that they think of the needs of their families. This is symbolized by their putting aside some leaves for Patna Devi (the goddess of the leaves) each time they go to collect fodder. These are small, perhaps invisible, but significant steps towards the recovery of the feminine principle in the forest. This recovery re-establishes the integration of forestry with food production and water management and it allows the possibility of a re-emergence of the diversity and integrity of life in the forest, of fauna and flora, of plants big and small, each crucial to the life of the forest, each valuable in itself, each having a right to participate in the democracy of the forest's life, and each contributing in invisible, unknown ways to all life. Diversity of living resources in the forest, natural or in an agro-ecosystem, is critical for soil and water conservation, for satisfying the diversity of needs of people who depend on the forest, and for the diversity of nature's needs in reproducing herself.

Chapter Fourteen

Co-operative Alternatives in Japan

Koyu Furusawa

At first glance, Japan appears to be enjoying a period of unprecedented prosperity, with its consumer boom and its new status as a financial superpower. It seems likely that the most important development in the international economy will be the global expansion of multi-national corporations as they spread their influence over what was until recently the Communist world. For industry in general, this will mean international restructuring to suit the demand for cheap resources and an efficient labour force, deepening the global division of labour, and world-wide corporate integration. Japan, which in not much more than 10 years has put itself well along the road to domination of the world economy, will be one of the axes on which this new global order will revolve.

In terms of agriculture and food supply, the consequences of these international changes will be that South-east Asia and other less-developed parts of the world will become the food bases to supply the industrialized countries. The benefits of Japan's economic boom do not extend to its agriculture, where there is a considerable crisis, most clearly shown by the rapid advance of events over the past few years concerning the importation of rice—the so-called 'sacred precinct' of Japanese life. If the importation of rice were to be liberalized, Japanese agriculture would receive a crushing blow. Many observers are already lamenting what they see as the coming extinction of the 2000-year rice culture of Japan, the country also known as the Land of the Vigorous Rice Plants. Many appeals are made in defence of rice-paddy ecology, in which natural dams hold great amounts of water, and of the rural economy, in which rice production, processing and distribution play a central role.

There are clear indicators of the severity of the agricultural crisis, but alternative activities are developing as a response to it, concerned with organic cultivation, ecology, and collective and co-operative movements. Before we look at some of the most significant of these, it is necessary to consider the background to the present circumstances, with particular reference to organic methods of farming.

The agricultural crisis now engulfing Japan has led some researchers and activists in Japan to look again at traditional agriculture and draw lessons from the experience of Japanese farmers over the centuries. In olden times and up to the early post-war era, farmers brought their produce to local markets early in the morning for consumers to buy directly. The average farm was small and self-sufficient, and farmers managed their land according to a composite approach in which a variety of purposes were realized. Their approach was ecological, in that the self-sustaining capacity of the land was nurtured and an organic connection between farming and their personal lifestyle on the one hand and the house's surrounding environment on the other was maintained.

This pattern of organic farming can be illustrated by examining the actual situation of a farm in central Japan, which operated from the pre-war to the early post-war period. Its food production centred on the rice paddy crop supplemented by vegetables. A small number of domestic animals were kept and a few cash crops were grown. From the surrounding environment fertilizer, food, and water supply were utilized for various purposes in order to preserve the natural balance. The forest provided fuel for furnace and fireplace and for heat and cooking, with the ashes being used as fertilizer. It also served as a protection against wind and fire.

Deep in the mountains far from the house, cedars and other trees were planted. Their wood was then used for rebuilding the house or for making tools. In the nearer mountains, broad-leafed trees, such as various types of oak, were grown. Their fallen leaves were used as fertilizer and their thin branches provided firewood or else were left for 10 years, after which they became like charcoal. This form of charcoal was not only easy to use, but served as a self-sufficient source of fuel and could be sold for cash. Additional food, such as chestnuts and peaches, was also grown in the forest. The bamboo grove was important, since it was the source of bamboo shoots for eating and of the material from which baskets and cages were made, as well as a wide range of other implements for production and daily life. In the ponds and rivers, small fish and river prawns supplied animal protein. The stable supply of water between the mountains in the region provided the power for moving the watermill, which was attached to wooden equipment that ground the wheat into fine powder, cleaned the rice, or else made incense from the raw material of the leaves from cedar trees.

Before World War II, not only did organic agriculture flourish, but there were also interesting experiments carried on within what can be called an organic farming movement. After the war, the number of such experiments greatly increased within the context of a vigorous movement.

In 1961, however, the Basic Agriculture Law was passed, which promoted the modernization of Japanese agriculture. Production became specialized and monocultural as the Government gave support to a one-product-for-one-region approach that led to the creation of a large-scale economy. The Japanese organic farming movement that had existed since pre-war days then entered into decline. During the period of high economic growth, the incomes of those not working in agriculture rose far more quickly than those of farmers. In order to catch up with average living standards, farmers had to switch over to large-scale, higher-efficiency monoculture. Another alternative was use of greenhouses for intensive production of high-priced vegetables.

The increasing dependence of farmers on large-scale distribution was another contributor to the simplification and concentration of production. Farms became larger—though not as large as those in Europe or the United States—and farmers used machinery to cultivate intensively a single crop. Chemical fertilizers came into widespread use as the application of chemistry to farming advanced. The use of pesticides led to sudden increases in production. A small application of chemicals could protect from the threat of blight and insects the crops of those who wanted to grow more and raise productivity. Not surprisingly, the previous way of doing agriculture came to be viewed as old-fashioned and backward, since it involved long hours of manual labour, tilling the earth, and putting in barnyard manure. Interest in the old ways did not resurface until the severe pollution problems of the late 1960s and early 1970s, at which time organic agriculture began its comeback.

In pre-modern times (before 1868), 80 per cent of the people worked the land and citizens' movements did not exist. Ordinary citizens' buying groups and consumer unions began during the period of inflation after World War I, when the high price of rice led to riots, but these groups did not last. When Japan geared up its war economy in the 1930s, the consumer movement was suppressed and destroyed. The next phase of the consumer movement was in the late 1960s, as a response to threats to life and health caused by pollution. The mass production and mass transportation of food had been consolidated and accompanied by the introduction of food additives and the use of agricultural chemicals. Movements arose in response to this trend, and their diversity and the intense level of their activity make Japan something of an experimental laboratory in this area.[1] Let us have a look in particular at the Seikatsu Club[2] and the new wave co-operatives, at the Society for Reflecting on the Throwaway Age, and at Masanobu Fukuoka's 'Do-Nothing' Farming.[3]

The Seikatsu Club, winner of a Right Livelihood Award in 1989, is the largest food-buying co-operative system in Japan, and a significant part of the organic farming movement. It involves 11 co-operatives and has a household membership of 180,000 spread through central and northern Japan, with total capital of nearly nine billion yen. In 1965, a housewife in Tokyo decided to find a way to avoid the high price of milk charged to consumers. Her idea was to band together with 20 other consumers in her neighbourhood and buy milk directly from distributors. This was the beginning of the Seikatsu Club, and over the next few years clothes, cosmetics and food also came to be purchased wholesale. By 1968, the group was big enough to form the Seikatsu Co-operative Union. In 1971, club members began to deal directly with farmers and take care of distribution themselves. The following year, agreements with farmers were achieved for rice, meat and fish. Two years later, members began ordering soap powder to replace the synthetic detergents that were polluting lakes and rivers. In 1978, a new headquarters was established in Setagaya ward, Tokyo, and the first Seikatsu Club housewife was elected to local government in 1979. There are now 31 members of the Club holding elective positions in local government in the Kanto area, central Japan. With the slogan 'Political Reform from the Kitchen' as an attention-getter they have successfully appealed to public concern over issues of safe food, preservation of nature, women's rights, peace, and grassroots democracy.

Unlike other consumer groups, the Seikatsu Club has extended itself to the national level and entered politics, so that it has the numbers and the organizational resources to influence Japanese society in a fundamental way. It is the coexistence of a small number of ecological activists with a large number of housewives concerned about getting cheaper and safer food, in one organization, that is the strength of the Club. Great numbers of organized consumers mean lower unit prices and also ensure that producers will link up with the Club, since a large, constant consumer market is available. At present, however, the Seikatsu Club faces something of an identity crisis. Its large-scale operation has provided benefits to conservative housewives, which the smaller groups could not match, but with the boom in organic farming and the entry of the large department stores into this market, the Club loses out to interests whose scale of operation is larger than its own.

On the other hand, consumers of more advanced ecological consciousness prefer the smaller, new-wave groups, which compromise less. It appears that the Seikatsu Club must develop a new strategy or it will lose influence within the co-operative movement. The most striking aspect of the outlook of some new-wave co-operative members is the importance attached to the relationship between producer and consumer. This relationship is sometimes spoken of with reverence and considered far more important than the mere pursuit of safe food to consume. Consumer members can go to the farm which produces their food, get to know the farmers, and find out how they live. They often organize the distribution of their own food and take part in the weighing and allocating of other food which is distributed. When eating, they may imagine the circumstances of production and even the faces of the producers, while recalling the climate and natural features of the area from which the food came. Such acts may seem superfluous from the standpoint of economic rationality, where consumption is defined as merely the process of using a physical object. But if consumption is reinterpreted as the richness of the production process within the context of a deep connection with the producers, their fields and the wider natural environment, then it becomes an act that realizes a wide plurality of values.

The farmers, too, gain from their relations with the consumers. They appreciate their own contribution to the well-being of consumers with whom they have personal contact and overcome the common tendency to downgrade their occupation. Here, the urban consumers play an important role in that their wish to live among nature and their vivid descriptions of the contradictions of city life make city life seem less attractive to the farmers. At the same time, such attitudes help reduce whatever gaps in consciousness have existed between rural producers and urban consumers.

Since producer–consumer co-operatives skirt market mechanisms, both producers and consumers must shift their way of perceiving economic relations. The consumer must recognize that, since organic farming is a method of farming dependent on nature, it is highly subject to weather and climatic conditions. Imbalances resulting from over-production and insufficient production easily occur and, what is more, since the growing of various crops is limited by season, the period in which it is possible to supply a particular type of food is correspondingly limited. In this miniature food-control system, the consumer may not always receive the desired amount of produce and cannot eat certain foods out of season. In addition, the time and effort put into cultivation increase in the absence of agricultural chemicals, which in turn is reflected in higher prices. This situation must be understood as the result of producers' efforts to supply safe food.

Prices are set through direct consultation among the parties involved, which gives consumers a better chance to understand the conditions of production. In this system of consultation, the fluctuating market price is largely ignored except in cases of extreme divergence from the set price, whereupon an adjustment can be made if any side so requests. Twice a year a plan is drawn up concerning what and how much shall be grown. Such consultations can be the basis of a fulfilling and enduring partnership only if there is a close connection between producers and consumers. When mutual understanding is absent, the relationship will break down. The important point here is education. New-wave co-operatives have no regular programmes to inform consumers of farmers' concerns, but when one enters such co-operatives, the terms and conditions are carefully explained. The Society for Reflecting on the Throwaway Age has classes for learning farming, including the planting and growing of rice, the general cultivation of plants, fruit, and flowers, and the processing of agricultural products. Such activities help raise the level of awareness concerning agriculture among urban residents.

The Society for Reflecting on the Throwaway Age was created six months before the first oil crisis in 1973, by Tsuchida Takashi and several others in Kyoto. Tsuchida was teaching in the engineering department of Kyoto University and had become convinced that the petroleum-based civilization could not last indefinitely. His critique equated the treatment of things—quickly used and quickly disposed of—with the treatment meted out to other human beings and the Earth itself. Things and people that are of use are exploited; when they are no longer of use they are cast aside. Such thoughtlessness ultimately threatens the survival of the human race.

The movement was first involved in paper-recycling, and then became concerned about food safety. Forming direct links between consumers and producers to ensure a supply of uncontaminated food was considered insufficient: production, distribution and consumption had to be viewed as a unitary phenomenon, so that the interests of producers and consumers could be identified with each other. A logical approach seemed to be to involve consumers in production. Farmland was rented and some producer members helped manage it while consumers performed the basic labour. Then the Society took over the accounting. The next stage was for the Society to purchase its own fields, with the consumers managing all the production and distribution activities. By working in the fields themselves, consumers learned that production and consumption are a unity.

Relations among consumers themselves were the next focus. Agricultural products are delivered with dirt on them to the residents of a particular neighbourhood, who must then divide them up among themselves. At such times there is opportunity to talk about the taste of food and ways to prepare it, which helps to restore vanishing relations among neighbours. An experimental farm serves as a summer camp and agricultural school for children.

There are now 1,800 members in the Society, including 80 farmers and a full-time staff of 10. The Safe Agricultural Produce Supply Centre functions as an agricultural relations centre that enables distribution to proceed smoothly. There is an agricultural foundation sponsoring organic farming, and the Yamagishi system of poultry-raising—an organically-based farming method—is promoted. The Supply Centre is funded by members' investment and run along the lines of a workers' collective.

Although co-operative buying of organic agricultural products has been criticized as expensive, it has the capacity to make people's lives less expensive. Changes in diet reduce overall food and household expenses, as needless food and goods are no longer purchased. Participants in co-operative systems feel that they have greater self-sufficiency and improved health.

For many non-Japanese, Masanobu Fukuoka epitomizes the ecological consciousness contained in the Japanese tradition. His image is that of the Eastern sage following the way of nature, who conveys Japan's ecological and spiritual message to the world. Some foreign people have even come to Japan expecting to find Fukuoka leading the organic farming movement. (Strictly speaking, Fukuoka denies that he is an organic farmer, since organic farming is based on scientific theory and Fukuoka provocatively disclaims allegiance to any theory.) But the reality is quite different. Fukuoka is not known by many Japanese and his influence in Japan has been small.

Fukuoka was a research scientist immersed in contemporary thought who went off on his own intellectual pilgrimage. He rejected scientism and wound up at the source of Lao Tzu's thought. With a philosophy of nothingness derived from his mystical experiences he applied himself to agriculture, giving his vision concrete form by creating his own world at his family's farm in Shikoku. Since his way of farming involves much less human intervention than is normally found in Japanese intensive cultivation, it has been called 'do-nothing' or 'natural' farming. It seems that, despite the attractiveness of Fukuoka's ideas, the results have been only fair. Some fields have outstanding harvests, while the yields of others are mediocre. It is a continual process of trial and error. He has had no effect on surrounding farmers. Nevertheless, given the total post-war retreat of the Japanese from all serious forms of spiritual thought, his philosophical ideas are extremely important.

In Japan before World War II, there was an attempt to pursue modernization while preserving traditional social attitudes. This hybrid has been called 'Western knowledge, Japanese spirit', and it led to the promotion of spiritual philosophies opposed to western materialism. But the entanglement of spirituality in Japan with nationalist ideas and the Emperor system led to its total discredit after the war. Japanese post-war thought has been materialist and pragmatic, and science has been worshipped as the source of power and prosperity. Fukuoka's uncompromising stand against modernity has made him an isolated figure, but a reaction against scientism in Japan is just beginning.

One should not focus solely on Fukuoka's conclusion, nothingness or emptiness. This goal of much Eastern spirituality, including Zen, has been criticized by many socially-conscious people as nihilist or life-negating, but what should be kept in mind is that such a conclusion comes at the end of a very long process. If the process and its dynamism are understood, then the value of a spiritual approach like Fukuoka's can be appreciated.

Humans and nature are most closely connected in the realm of agriculture. Nature exhibits power and humans live within this power. Nature and humans interact and interrelate; through such intervention nature may break down or it may be improved. When it is improved, humans also improve, as nature and humanity relate dynamically and reciprocally. It is this dynamism which will call forth a response from those with inner conflicts. Since people search for self-realization through external pursuits in today's civilization, such dynamism is not widely understood and so people like Fukuoka, who are exploring the inner dimensions, are viewed—especially in Japan—as inhabiting a totally alien world.

This inner world and the connection between humans and nature does not, however, exhaust reality. Human beings are involved with production, distribution, community, society, economics and exchange, human relations and culture. So, although the relation with nature is the source and extreme limit of being, concern for the realm of interpersonal relations and society is an essential part of living. This area is one which Fukuoka has ignored. The concept of organic farming goes beyond questions of production; it also goes beyond responding to an individual's mental anguish. Though the latter is not to be made light of, a philosophy that can lead to the creation of new social relations is also required.

The force behind the concerted action by local residents and the consumer movements of the 1970s can be expressed through words such as 'safety' and 'health', and put into action from viewpoints such as 'locality', 'livelihood', and 'quality of life'. After entering the 1980s, though, these viewpoints were incorporated into corporate strategy and led to a health food boom. 'No additive' foods and organic agricultural products began appearing in supermarkets as luxury items around this time. This trend should be understood in terms of Japan's evolution as a consumer society. As Japan has reached the apex of consumer culture, food from all over the world is now available there. In consequence, people have begun looking for products with a difference, something that is natural and hand-made, with little concern over price. Such desires make consumers easy prey of major commercial capital. Along these lines, it seems that organic farming is increasing its commercial colouring instead of remaining a straight farmer-and-consumer movement.

There are bound to be pitfalls in such a trend. Much produce in the central Tokyo commercial market is labelled 'organic', even though only a few organically-grown items have been thrown in among food grown with chemicals. A further disadvantage is that the presenting of organic farm products under a brand-name jacks up the price for the safety and authenticity which are intrinsic features of the products. In such ways a meaningful relationship, which is supposed to benefit producers and consumers, becomes a mere business transaction in which both end up as losers.

At present, the organic farming movement is discussing the question of formulating standards for determining whether or not produce can be deemed organic. The Japan Organic Farming Society is against the establishment of such standards, since it prefers to emphasize human relations of trust, which are possible in the co-operative movement. Yet consumers who purchase on the market do want assurance that they are actually buying organically-grown food.

More important than the issue of standards, is finding a way to form an alternative economy. The appearance of workers' collectives and the broader social involvement of groups centred around co-operative buying indicate that the first steps toward the forming of a counter-economy are being taken. This is the only direction open to the organic farming movement if it is to retain its impetus and resist the commercial forces capitalizing on the interest in organically-grown products.

The next step is for the meaning of work and the need for a meaningful life to become issues of widespread public concern. Quality of life needs to be subject to questioning in terms of its relation to one's job as well as to the economic, political and social world. With regard to the co-operative buying of organic agricultural products, the driving force of the movement must be not only safety but the treasuring of heart-to-heart encounters.

The future of organic farming in Japan is intimately connected with the future of agriculture itself, which depends on a revival of interest in farming among Japanese people. Most farmers are older people, and the overwhelming majority of those in their working prime who farm also work at other jobs, with the larger part of their income being derived from that latter employment. Many farmers who desire to marry find it difficult to secure brides, because women find farm life unattractive. Among younger people in Japan there has not been much of a back-to-the-land movement, but recently there have been indications of a re-evaluation of attitudes towards farming, nature and the countryside. A new sympathy for farming has just started to develop. It stems not from a liking for old-style agriculture and its hard labour or for old-style villages and their exclusionary attitudes, but rather from a desire to return to nature and to search for one's lost humanity in the countryside instead of the cities. From such attitudes a new style of farming and new types of farming villages could come into existence.

With Japan's economic prosperity, the people's level of life has changed and, although they cannot be said to be agrarian scientists or environmentalists, little by little they are developing sympathy for such notions.The movements in this direction in the United States and Europe are stronger and somewhat different from those in Japan, but as an overall movement there is much that Japan and the West have in common—a reaction against the excesses of the technological society. Now I am getting into the realm of the future and of dreams, but the fact is that the Japanese industrial system within the global economy is faced with many large problems. How long will resources be available for Japanese companies to exploit? How long can these companies continue to degrade the environment both at home and abroad before the price becomes unacceptable?

A balance must be created between the industrial sector and the as yet small alternative movements. Many possibilities exist: protectionism; agrarianism; even eco-fascism. But the aim should not be to reverse history by completely negating science; nor should individual rights be ignored. Rather than aiming for total upheaval, I feel that the Japanese alternative movements will focus on limited targets. There will be the environmental movement, including the organic farming movement; the peace movement; the Third World network group, and so on, all facing in the same basic direction. It is still too soon to provide details concerning the nature of this future society. In Japan, the agenda has not yet been determined, though time is beginning to run out.

However, new initiatives are taking place. In February 1989, alternative trading in bananas began, between a development group in the Philippines and some Japanese co-operatives. In Tokyo, in April 1989, the Citizens' Bank began to support local businesses such as workers' co-operatives and, in November 1990, Women's World Banking followed suit. I think that these are small but hopeful signs of a trend towards an alternative economy.

Chapter Fifteen

Human Ecology and the Sustainable Society

Robert Waller

MONOCULTURAL SOCIETIES DO NOT EXIST in the civilized world; past cultures survive as strata beneath the dominant culture or partly interlaced with it. As the dominant culture proves inadequate to deal with the problems arising through human activities, other cultural elements emerge, either from the past or newly originated. In my view the organic movement, which has deep roots in the past as well as the present, is developing into Human Ecology to form a counter-culture to the dominant culture of philosophic rationalism.

The conventional modern citizen, proud of being rational, as a consequence of the influence of science and its pragmatic experimental mode of thinking, has tended to disparage the past and to speak as if life only became worth living in the 19th century after the majority of the people who worked on the land were freed from their bondage to the farm and the village and millions of them were able to migrate in railway trains to the superior freedom of city life and transform themselves into urban people. Life was no longer to be 'nasty, brutish and short'. The mechanization of the production of material goods had overcome—or would overcome—the curse of scarcity which had been the cause of so much human misery. This triumph of the machine, including the industrialization of agriculture, has been regarded as if it were a step forward in the evolution of mankind, an example of its unquenchable, forward-moving human spirit, not only in its search for self-improvement, but to establish equality and justice as well; it seemed that, without doubt, constant economic growth would enable democratic ideals to be put into practice without personal sacrifice or offence to Mammon. Henceforward, all problems would be simply economic problems. With scarcity conquered and a just and enterprising economy self-regulating, all could at last turn their minds to higher things, to art and religion and philosophy.

In the Victorian period it was thought unpatriotic, bad manners and subversive to draw attention to the poverty that accompanied increasing prosperity and to the many industrial cities that had become ungovernable pandemoniums. In America, Henry George could not find a publisher for his book *Progress and Poverty.* It was forbidden to doubt; it was almost an article of the Protestant faith that mechanization, industrialization and capitalism were part of God's plan to lead the Anglo-Saxon race to a new Eden of prosperity. The effects on the environment and on the character and soul of the industrialized individual—often referred to simply as a factory 'hand'—were assumed to be beneficial; education was improving our intellects and museums were increasing our knowledge. That industrialization was also inflating our pride, making us arrogant and aggressive and giving us the military and naval means to enforce our will on other nations and cultures on a world-wide scale, was considered an act of God, who willed that we should be the agents of His peace through material progress and Free Trade. Free Trade was not a matter of free choice but must, for their own good, be forcibly imposed on other nations (whose markets were of value to the West)—as in the case of China and Japan.

It was this 'imperialist' attitude to material progress that gave Marxism its universal appeal. We should not forget that when we relapse into self-righteousness. But Marxism was also the offspring of mechanization and industrialization, like capitalism, and the likelihood is that the cultural inadequacies of both will lead to the downfall of both.

Critics of our rational, scientific culture are usually told they yearn to restore the Middle Ages. Such critics, in my experience, apart from not having read Aristotle, Augustine or Aquinas—for which they may be perhaps pardoned—have never even looked contemplatively at a cathedral and thought about what it tells us of the people who built it. And what will the future think of the people who built the brutish, faceless, ugly commercial buildings of today? Certainly these buildings are only the tombstones of a class of our contemporaries with too much power and too little culture. Similarly, the Inquisition by no means tells us all about the Middle Ages; it was the legal arm of the authoritarian bureaucracy that finally brought about the downfall of the ecclesiastical regime as men rebelled against having their intellect, imagination and daily life regulated by dogmas that increasingly appeared unrelated to their real Faith. The political exploitation of the Inquisition became more oppressive as the Church and its guardian, the Spanish Empire, felt their authority increasingly challenged by free thought.

We should not forget that the astounding width and power of Shakespeare's integrated imagination and intellect were due in part to being close to medieval Christianity as well as being appreciative of the Renaissance and the exploratory spirit of the Baconians. Our own great need is to recover such a diversity of interrelated interests and transcend our growing tendency to a monocultural Philistinism. From a broader perspective, what will our civilization look like in 150 years' time, with the massive scale of our wars, our nationalistic, racial and religious hatreds, our devastation of nature, pollution, economic exploitation through free trade and high interest rates on loans to those poorer than ourselves...? Will a city with 2000 murders a year like New York be held up as an exemplary paradise that has overcome scarcity and justly distributed the fruits of progress? To say the least, it will not look all that superior to other ages. Most of the earlier periods of culture had a much greater sensitivity to beauty, aural and visual—Shakespeare, being a genius of his time, was as much concerned with the euphony of what he wrote as with its meaning. Beauty was felt to be a mystical proof of the existence of the gods.

We may well appear in the history books as 'The Age of Mammon'—the Mammonites and Millionairians—a consequence of the industrial revolution not predicted by its democratic, rational idealists.

Contemplating history in this way must surely lead us to conclude that we ought to assess every age on its own account and not as only a stepping stone or transition to an even more 'evolved' culture—such as the 'superior' civilization of our own time. The concept of evolution has misled our thinking in this respect. Civilizations and cultures do not evolve in the way that natural organisms have evolved over millions of years. Human evolution has been complete for many thousands of years now and our 'evolution'—if we can use the word metaphorically—is the consequence of our cultural achievements which are embodied in our inherited works of philosophy, art, science and learning. We can, in this respect, regress as well as progress. And disparaging the past en bloc, as it were, is a form of regression.

To disengage from the cultural domination of Mammon will not be easy, for its values are now so much part of the very air we breathe they seem like nature itself.

Somehow, we must find a way of conjoining the best of the past and the present. This I consider the true conservatism and a pathway open to any political party, though it is non-governmental organizations that are nearest to it; those, for example, that strive to conserve the environment, to protect nature against the ravages of ill-considered development, keep alive the achievements of the cultures of the past and encourage creative talent.

What we need is a comprehensive and philosophically based framework within which we can assess how decisions ought to be made in harmony with a sustainable culture and society. This is being created by the Human Ecology movement.

Everything has its ecology which is provided by nature, which itself is sustained by the cosmos.

Human Ecology is essentially holistic, which means in practice that we should not be content to think with the logic of the Roman road in a straight line—such as 'Whatever makes the most profit is best and right'—but must consider, as far as is possible with present knowledge, other relationships—effects on employment, the environment, health and safety, aesthetic and cultural values, conservation of energy and resources... The human ecologist must remind himself, even after he has done his best, that 'Whatever we know, there is always more to be known'. This need not be depressing; it can make life more interesting.

Human Ecology starts with the person, insofar as the nature of the knower plays a part in what is known. The knower must, therefore, pay attention to his own development. This is one reason why the young declare a preference for the arts rather than the sciences—to the dismay of Governments, they think the arts will give them more self-knowledge. The concept of objective knowledge that has prevailed during our mechanistic era—the cosmos as a machine—has neglected the part played in understanding by the personality and beliefs of the observer or learner. For example, the concept of 'the survival of the fittest' may, unconsciously, be an adaptation of what Darwin called 'natural selection' to provide the sanction of nature for the prevailing competitive economics. Another destructive consequence of the belief in absolute objectivity is the assumption that technology is neutral, whereas nearly all Western technology has been designed to advance the self-interest of its makers. In a different culture the technology would be different. 'Ah,' says the objectivist, 'there can be only one successful and true culture, that which harmonizes with our objective knowledge.' A chain of catastrophes has exposed the limitations and falsifications of this attitude to knowledge and to what extent it is based upon false assumptions about nature that have immodestly reduced our concern for human error. The human ecologist, taking into account all these factors, learns that disappointment and disillusion may serve as great teachers, and even inspire those with humility to new effort and greater work. The beginning of wisdom is to be sure that we are unlikely to be exactly right and that we cannot always have our own way. To a certain extent this experience underlies the principles of 'uncertainty' and 'probability' in modern scientific thought which have displaced the former confidence in discovering laws of nature that have no exceptions.

If we now consider a major principle of human ecologists this will give us an idea of its practical effectiveness and how it has incorporated and extended the insights of the organic movement.

This is *carrying capacity*. The value of this concept is dramatically illustrated in the feud between the agroforesters and the ranchers in the tropical rainforest regions. The ranchers require about 5 acres to keep one animal on their bleak grasslands. This is because, in this environment, grass fails to protect the soil against the heat of the tropical sun and the torrential rain—though in other climates it may be the best cover. The Indians have known for thousands of years that trees alone serve this purpose of protecting, sustaining and nourishing the structure and fertility of the soil. Some of the older settlers know this too and, like the Indians, they combine farming and forestry—agroforestry as we now call it. Five acres of agroforestry can sustain a variety of economically valuable trees bearing exotic fruits now sold on the European and Asian markets. These trees guard the soil and draw up underground sources of water with their deep roots. Between the trees, root crops can be grown. Agroforestry is in harmony with the forest ecosystem, i.e. interrelations. Its sustainable carrying capacity is far greater than that of the ranchers' grasslands—which in any case have only a limited life before they are worn out and become useless until an agroforester takes them over. They can then be gradually restored to fertility. The greatest evil of all is, of course, logging.

Not only does agroforestry have all these biological virtues, but it provides homesteads for many families which, otherwise denied access to land for establishing themselves with a sustainable agriculture, resort to destructive practices for a livelihood, such as polluting the rivers with mercury in pursuit of gold. It is hard to believe that the ranchers and loggers are prepared to murder people to keep them off the land their own practices ruin, but the evidence is overwhelming; we have even seen it on films shown on the television by adventurous conservationists.

Although the ranches belong to brutal landowners with a long tradition of oppressing the poor, nevertheless we can, I believe, have a reasonable faith that when their evil practices are universally exposed, there will be reforms. But, for these to be successful, people, world-wide, must learn that there are alternative practices that are ecologically and economically superior, as is shown by applying the concept of carrying capacity. This is a concept of human ecology, because it includes human beings as well as plants and animals and husbandry. It is an integrating concept based on the *interrelationships* of all these factors. Human ecology is the foundation of a new culture.

The diffusion of the right kind of knowledge is, therefore, another essential for the survival of mankind on the planet. An educational network of distance learning is needed such as has been technologically possible only in the last decades. This has been started by the Commonwealth under the direction of James Maraj, who is also president of the Commonwealth Human Ecology Council. The failure of the world banking system to relieve poverty and conserve the environment shows the need for a human ecological framework within which economics can function an 'ecolomics'.

The Commonwealth Human Ecology Council, as far as I know, is the only organization that actually attaches human ecology firmly to its masthead. It is a pioneering organization founded some 20 years ago by a remarkable New Zealander, Mrs Zena Daysh, who still tours the world confronting prime ministers and officials with the need to include human ecology in their thinking and planning, and who is becoming so influential that CHEC is now officially attached to the Commonwealth Prime Ministers' Conference (CHOGM). With very little money and with a dedicated enthusiasm, Mrs Daysh and her supporters have established a world-wide network of CHEC groups and projects. The Council operates from a basement room in London. Two years ago they held a conference in Edinburgh to discuss education in human ecology and what effect it should have on every aspect of life. Participants from 47 nations took part, including some from Eastern Europe and China as well as the Commonwealth countries. Relations with American human ecologists have been close for a long time. Many universities world-wide are now setting up human ecology departments and organizing human ecology conferences attended by CHEC. One American college, the College of the Atlantic, gives degrees only in human ecology. In India, the University of Rajasthan has a large human ecology department and has set up a CHEC for India. The President of Guyana gave the opening address at the CHEC Edinburgh Conference and said that 'After Mrs Daysh's visit Guyana was never the same again'. He has since donated to the Commonwealth one million acres of rainforest for ecological management. UNESCO is to publish a book of the proceedings of the CHEC conference—its ninth international conference—chosen from more than 100 papers submitted. Mrs Daysh now visits Japan and has been given funds and support by a Japanese Buddhist movement.

This brings me to my final point. CHEC recognizes, naturally, that innumerable voluntary non-governmental organizations concerned with the conservation of the environment and wildlife and the organic farming movement, and many community movements and peace movements now constitute the spearhead of a new culture which, they believe, can be integrated in the philosophy of human ecology. These allies appear all over the place—often in religious, artistic and scientific groups who are aware of the need for a change of culture.

CHEC respects the cultures and religions of the past. It realizes that in many countries in Asia a renewal and reinspiration of the teachings of the Buddhist and Hindu sages may be the most potent force in saving the environment from the onslaughts of western commercialism and the contemporary western culture with its sterile mixture of positivism, utilitarianism and pragmatism which leads people to despise their older cultures and religions for keeping them poor and under-developed—under-developed meaning 'not westernized'. These ancient cultures and religions have taken thousands of years to develop and western influences can weaken them in a few generations. East and West should, of course, learn from each other, but we must be aware of the vast corrupting temptations offered by western commercialism to enrich ourselves at the expense of the environment, of the poor, and of non-renewable natural resources.

However, in the Indian settlement of Auroville, efforts are being made by an international community to tackle economic and social problems by methods in harmony with eastern teaching—in the spirit of Schumacher's *Buddhist Economics*. In the Chipko movement, forest communities are chaining themselves to the trees to save them from the loggers and to rebuke the folly of the Nepalese government in permitting the decimation of the forests. In Sarawak similar protesters have been arrested.

The example of the Chipko movement and many other protest movements in which the poor, at the risk of imprisonment or even death, defend their homes and their traditional environment from the destructive actions of western commercialism and the western style of self-enrichment, is a sign that if anyone inherits the Earth at all it will be the poor. Until we recognize the rights of the poor to their own land and culture, and cease to listen to the serpent of economic rationalism, we shall expel ourselves not only from our prosperous Eden, but from the planet itself.

For the contemporary rational positivist it is 'emotionalism' to call anything a struggle between good and evil; nevertheless, from the old fashioned ethical point of view, this is certainly what it is. It is not enough to say that evolution or reason or simply time will put it all right; something else is required which used to be called redemption, the confession of sin. The most apt of the parables to our own time is that of the Prodigal Son. Similarly, Krishna understood the spiritual unity of nature and the gods.

Human ecology may take a long time to change our culture, but it is bound to do so in due course, because it has a suppressed human potential—the soul—as its greatest ally.

Chapter Sixteen

Towards an Organic Energy Policy

Diana Schumacher

'Energy used above a certain level both pollutes and destroys'—Ivan Illich, *Energy & Equity* (1979).

THE USE AND ABUSE OF ENERGY since the industrial revolution has grown and intensified exponentially and currently involves the entire planet and the biosphere in the greatest feat of pollution and destruction ever perpetrated by that small and dangerous species, homo arrogans. We are now in a global pollution crisis of such magnitude that it cannot be solved by more disparate measures, more technical, political or economic fixes, more conferences or policy statements, more misdirected capital investment or even more research. There is now needed an *integrative perspective* which looks at the whole and not just the parts of an issue. The requirement is for a 'systems level' or holistic understanding of energy.

Energy fuels every aspect of our everyday lives and the 'energy problem' cannot be isolated from other realms of economic, social and political affairs. Sustainable policies must at last recognize that the 3Es of Energy, Economics and the Environment are inextricably interconnected, and we ignore these connections at our peril. Also required is a further practical set of 3 Es—Education, Encouragement and Example. What is demanded most urgently is a Global Green Agenda which lays down certain fundamental principles of reconstruction that can be applied simultaneously to all the interrelated areas of biological destruction and poverty, including energy poverty—a new phenomenon of this century. This chapter suggests *ten principles* or *energy commandments* on which long-term policies could be based and describes what some of these organic policies might be, illustrating them with a few practical examples.

Ten Energy Principles (or Ten Commandments)

1. The Principle of Efficiency.
2. The Principle of Right Scale.
3. The Principle of Empowerment.
4. The Principle of Co-ordination.
5. The Principle of Diversification.
6. The Principle of Flexible Integration.
7. The Principle of Low Risk.
8. The Principle of Evaluation.
9. The Principle of Social Justice.
10. The Principle of Sustainability and Replenishment.

These ten 'commandments' can act not only as guidelines for governments, but may also activate initiatives at local level, both individual and collective, and form a powerful framework for action.

The Ten Commandments can also serve as reference points against which policy investment decisions may be set. Together, they form a systemic or holistic basis for resolving many of the seemingly intractable and interrelated global problems.

1. The Principle of Efficiency

In energy terms, the principle of efficiency means that the most favourable balance possible must be struck between energy inputs and total energy outputs.

Efficiency is concerned with maximizing the output from any given input. It is, therefore, imperative for conservation and the elimination of waste, including the avoidance of harmful by-products.

Efficiency also ensures that the maximum value is added in the process of energy conversion. Only in this way will the Earth's scarce resources be husbanded carefully and not squandered.

Since the 1973 Middle East crisis there have been numerous academic researchers and Government organizations *measuring* the effectiveness of public policy in promoting improvement in energy efficiency. Both public perception of the problems and energy efficiency have increased within limited boundaries. To quote even the Brundtland report of the World Commission on Environment and Development (1987), 'a safe and sustainable energy policy is crucial... Energy efficiency policies must be at the cutting edge of national energy strategies for sustainable development'.[1] What does this overdue rationalization mean in real terms, and above all the term 'sustainable development'?

The International Energy Agency (IEA) report published by the OECD (1988) points out that, although in the developed countries there has been considerable progress in the rational use of energy over the previous decade, particularly in the industrial sector, there has also been reduced public interest, due to more abundant energy supplies and the lower energy prices prevailing in real terms. It admits that 'greater energy efficiency extends the availability of depletable low-cost fossil fuel reserves and therefore reduces the need for additional supplies which might in future come under increasing pressure for security, environmental or economic reasons'.[2] But the same report acknowledges that, although there have been gains in fuel efficiency, there has been a steady increase in demand for transport fuels *'due to other factors'*. Undoubtedly these factors include the increased number of air and motor transport vehicles which government policies continue to allow, and the fact that pricing policies nowhere reflect the true cost of environmental degradation or depletion of reserves. In other words, the principle of energy efficiency also has political and educational implications and is not just an economic issue.

Apart from transport and transport systems, one of the main areas where energy efficiency can be increased is in *improving building stock*. This is a certain way of safeguarding and maintaining energy supplies for future generations. Unfortunately, in the industrial countries of Western Europe, good building practice in the energy sense has been lacking. Passive solar buildings such as those recommended by Archimedes and constructed by the ancient Greeks and other conserving civilizations, are specifically designed to trap the sun's rays for heating the interior of buildings. Modern examples now exist throughout Europe, the United States and elsewhere and they hold an important key to our energy-efficient future.

By the simple device of introducing better building insulation and conservation practices, say to the same standards as the Danes (who have the highest standards in Europe), Britain's production of carbon-dioxide (CO_2) in power stations would drop to a third of the present levels; from 65 million tonnes per annum to 21 million tonnes.[3] On a small but nevertheless significant scale, most electrical appliances from light bulbs to water heaters may be cost-effectively redesigned to require only one-third of their present energy consumption, which could contribute a considerable energy saving almost immediately.

The conversion of primary energy to useful energy at power stations and the subsequent transmission of power to the consumer, often over great distances, at present accounts for approximately another third of energy loss and unnecessary CO_2 production. One can, therefore, see how great is the cumulative energy saving by having overall energy-efficient systems. In this respect, investment in combined heat and power (CHP) and district heating systems is of utmost importance. Such schemes have already proved cost-effective over many years in different countries and under very different circumstances from Denmark, Jamaica, Germany and more recently the UK. Energy saving throughout the agricultural and agrochemical industries could also be very considerable by switching to less energy-intensive and more energy-efficient methods of food production. Efficiency also includes the questioning and revaluation of necessity. The petrochemical industry and other heavy industries, such as the manufacturers of steel and aluminium, are among the most energy-demanding, but not all their products are by any means necessary in an ecologically stable society.

Another step in energy efficiency with implications for the building, engineering and manufacturing professions as well as for all consumers, lies in the 4 Rs—reclamation, recycling, repair and re-use. Energy Efficiency rejects the wasteful practices of a throwaway society and examines the energy savings to be gained by encouraging the recycling habits of every generation throughout history—excepting the present one.[4] Where recycling is simply not possible, methods are being explored to find the most efficient and environmentally-benign methods of producing fuels from domestic and industrial wastes through landfill gas recovery, incineration and heat recovery for electricity generation, pyrolysis, waste-derived fuels and fluid-bed refuse incineration.[5] All waste is simply a misplaced energy resource in one form or another.

Taking such methods of energy-saving into account, the World Resources Institute Report, *End Use Global Energy Product* (EUGEP) in 1990 affirmed the technical feasibility of halving industrialized countries' energy consumption by the year 2020. A report by the University of Princeton, published in 1990, reached similar conclusions. Also in 1990, a report submitted by the Department of the Environment to the Intergovernmental Panel on Climate Change, the UN sponsored body investigating global responses to the greenhouse effect, suggested that Britain's energy use could be cut by as much as 60 per cent over the next 15 years through energy conservation measures alone. The research and evidence produced during the last decade by world-wide establishment bodies now finally endorses the pioneering work of Gerald Leach, Amory Lovins, Friends of the Earth et al. a decade earlier—that energy efficiency is the most practical, holistic and cost-effective way to save energy and curb greenhouse emissions simultaneously.

2. The Principle of Right Scale

Energy supply systems should be designed to serve local communities of the maximum economically self-sustainable size and be appropriately matched in scale to the size of those communities.

The first energy principle is basic since it relates to all the rest and all must overlap. *The second principle emphasizes that the human being and the natural world should always remain visibly at the centre of human endeavour and organization.* It rejects giganticism and the sense of isolation, dependence and the diminishment of human dignity which this brings.

Each local community should make use of energy sources within its own area and strive, as far as possible, to be energy-independent from other areas. Ideally, the size of these independent areas should be commensurate with the ability of the population to be self-reliant in other key aspects of life, such as agriculture, basic manufacturing, primary and secondary education, primary health care and local government. Hence, the scale of energy supply systems will vary according to the geographic locations, demographic profiles, and needs and possibilities of those communities they serve.

A good example of an energy plan that is compatible with the principle of right scale is the Cornwall Energy Project in the most southern county of the UK. Proposals contained in this 10-year Energy Action Plan show how Cornwall could reduce its fuel import bill by £42 million a year by adopting common-sense conservation measures and harnessing some of the County's abundant sources of renewable energy. At present, it is estimated that Cornwall is wasting more than £100,000 in energy each day since most of the energy is imported from areas further north. These areas not only have to bear the environmental and health costs, but are not in touch with the exact end-use requirements of Cornwall's locally varied needs.[6] Right scale ensures maximum energy efficiency and responsibility.

3. The Principle of Empowerment

The Empowerment Principle emphasizes the need for autonomy and self-determination in energy policy matters.

This seeks to ensure that there is maximum delegation of power and autonomy to the grass-roots, or at least to the lowest practical decision-taking level. Empowerment automatically creates an atmosphere of involvement. It follows that each community should, *as far as possible*, be responsible for conducting its own energy policies and practices.

The key phrase is 'as far as possible', since some dependency on wider national or international policies will be essential. On the other hand, the main thrust of organic policies should be determined locally, where knowledge of real needs and available supplies is the greatest.

To give a practical example, in most cases rural electrification schemes transmitted by grid from centralized power stations are entirely inappropriate for poor Third World communities, since only the rich landowners can afford connection to the grid, let alone the cost of the electricity to be used. Although, during the 1960s, the World Bank provided more than 25 per cent of its lending to rural electrification, today fewer than 25 per cent of rural households overall have a regular supply of electricity, the prices paid for electricity are often higher than in industrialized countries and the generation and transmission inefficiencies far greater.[7]

As long as energy supply decisions are left to officials in central governments or to the international agencies, such inappropriate policies will be perpetuated. With local control, more user-oriented supply systems, which can reach the poor, would be developed. Impressive pioneering examples, as well as the typical obstacles to such self-reliance transitions, have been documented throughout the Third World in such varied places as Sri Lanka, Nigeria and Colombia.[8]

There are also great dangers in energy dependency especially resulting from reliance on imported fuels on a large scale. As the Gulf crises have demonstrated, the entire world is vulnerable to political instabilities—and, as usual, the poorest suffer first and most. Only 'communities of resistance' with control over their energy supplies and management hold the key to sustainability and can maintain a dependable infrastructure. It requires a turn of the switch in our thinking from abstract notions of growth and development to see what policies are best for the real world.

Empowerment also involves education and facilitation, with particular regard to the role of women, who constitute 50 per cent of the population but who are mainly disregarded in energy-policy decisions. Women, nevertheless, are the main consumers of domestic energy, purchasers of household items and are increasingly becoming users of transport fuels, especially in industrial countries.

Maximum local control over energy policy for a community has many organic consequences. One is a much better ability to oversee the thermodynamics of matching energy supply and demand, thereby eliminating mismatches, spare capacity and waste. Energy demand forecasts are also more accurate if locally based. Details of present and future consumption and local supply patterns are more clearly understood and there is local understanding of and responsibility for the supply of local needs. Past attempts to forecast primary energy requirements at national or international levels have proved hopelessly inaccurate in most industrial countries, as is exemplified by the extremely variable estimates of the International Energy Agency and other authorities over the past two decades. This is partly because there has been an unsafe reliance upon abstract notions, such as the relationship between GDP and energy consumption ratios for a particular fuel or industry group. This has proved as irrelevant as linking GNP to the quality of life! Even the best mathematical models are, of course, only as good as the assumptions fed into them and the proclivities of the forecaster.

4. The Principle of Co-ordination

Energy policy in each community should be locally co-ordinated.

This fourth energy principle recognizes limits to the substantive positive changes that can be achieved without some collaboration and co-operation by local, regional and national authorities. To safeguard the principle of empowerment mentioned above, each level of organizational hierarchy should recognize the knowledge and authority of local people (in this case energy users) to define their own requirements and seek full co-operation with them.

In policy terms, a practical step in communities with an established infrastructure could be a local energy authority. This would be established in each region to co-ordinate policy and ensure co-operation.

At the national level, some rationalization of institutions representing different energy interests (in the UK there are at present some 70 or so of these) would also be beneficial, to ensure more constructive collaboration. One suggestion is for regional and national energy control agencies, to co-operate on priorities and lay down a framework for energy good practice by the local energy authorities. This might include the establishment and oversight of common standards, the carrying out of audits of local practice and the provision of central funding for research and energy development. What is truly uncooperative is the increasing competition among national and multinational fuel corporations to encourage and increase their energy markets.

5. The Principle of Diversification

Energy supply sources should be diversified to provide the best 'fit' between available energy sources and types of end use.

The fifth principle, of diversification, may be translated into the requirement that appropriate energy supply systems should be designed to serve diverse end uses and different communities in a sustainable way. For what purpose is a particular type of fuel to be used? Where? By whom and for how long? With what environmental consequences? Correct answers to these questions will avoid short-term economic traps. It will also avoid exporting inappropriate energy technologies to countries whose resources and requirements are very different from our own and whose infrastructure and technical know-how at a local level may be non-existent.

Implementing the principle of energy diversification will probably involve renewable energy hybrid systems replacing many of the conventional mono-systems of most industrial societies. A differentiated energy supply system is crucial for sustainable energy development, since all conventional sources of energy supply have very short-term futures unless they are radically modified.

In the Third World, single-fuel dependence is already crippling many countries, especially those with reliance on oil. In rural areas, where biomass in the form of fuelwood and animal wastes accounts for more than 90 per cent of energy consumption, wood is rapidly running out and becoming a scarce and costly resource, whilst both agricultural wastes and animal dung are urgently needed to protect the topsoil and to fertilize crops.

Since the most impoverished rural areas of the Third World will not have the means of developing substitute energy resources between now and the beginning of the next century, the most urgent priority is to help to diversify their supply by gradually introducing other renewable-energy devices, such as biodigesters, solar cookers, water pumps and solar stills, at community level. At the same time, it is essential to ensure that fuelwood for cooking is used as efficiently as possible by introducing locally-made cooking stoves to replace the wasteful and polluting traditional open fires, and to ensure that trees felled for fuelwood are replaced immediately.

Already, there are hundreds of examples of hybrid systems combining whatever energy sources are available at the community level. One encouraging example is at the Ladakh Development Centre in Leh, in the foothills of the Himalayas, where there are absolutely no local fuel resources other than dung, which is needed for fertilizer. A combination of active and passive solar systems for heating and cooking, together with wind pumps and biodigesters, demonstrate the possibility of self-sufficiency even in this harsh climate. A further successful community-based hybrid system is to be found at the Schumacher Centre being developed by Dr Mansur Hoda in Lucknow, India. All these use the same basic renewable-energy technologies tailored to meet local needs on a sustainable basis.

6. The Principle of Flexible Integration

The various elements of a system should be flexibly designed, but display a powerful degree of cohesiveness, both in relation to each other and to the wider community of which they are a part.

This principle is, of course, complementary to the fifth principle, of diversification. Flexibility of energy supply is particularly important and demands that people are never locked into a rigid or intransigent system, since nature and human societies are in a state of constant and ever accelerating change. Systems should be designed in such a way that communities have the freedom to respond to changes as and when they occur. Energy supply sources must be mutually interactive, so that fluctuation or the failure of one source or a sudden increase in demand may be appropriately recompensed.

In terms of a particular conventional fuel's possibilities, oil is the most flexible. Nuclear electricity is the least since it can supply only electricity from a centralized power source at high cost and considerable environmental risk. Also, as it is capable of providing only baseload supply, it is unable to respond to sudden demand oscillations.

Within the range of renewable energies, biomass is the most flexible since it can be used for combustion, electricity generation, biogas production and liquid fuels for transport as well as having a host of other by-products which could replace many of those supplied by petroleum, gas and coal. Biodigesters themselves are a very flexible energy converter since they can be built locally, can vary in size to meet local requirements, use a variety of inputs from animal, plant and human wastes, produce methane gas (which may also be used for electricity generation) and provide residual slurry to make an excellent nutrient-rich fertilizer.

Interactive, flexible energy systems are much more robust and secure than those based on single supply sources, even in industrial countries. California, for example, has very high levels of energy consumption both in industry and agriculture. It used to rely on the main fossil fuels together with nuclear energy, but now has a deliberate policy to develop and extend its use of renewable energies, especially solar and wind. California is one of the world leaders, together with Israel and Japan, in developing solar energy technologies, especially solar thermal electric generation and photovoltaic cells (electricity directly from sunlight). All of these forms of renewable energy employ small-scale mass-produced modular units which are extremely flexible, quick to construct and versatile, which makes them also useful for remote areas. The large-scale renewable-energy projects, such as tidal barrages, wave-power devices and dams, are capable of delivering greater amounts of concentrated power where the location is suitable, but suffer from the same lack of flexibility as conventional power plants.

Already, there are more than 300 megawatts of solar thermal electric plant in California's Mojave Desert alone. On California's wind farms in Altamont Pass there are now 7,500 small wind generators, mostly installed by the Pacific Gas and Electric Company. It was seen that a system of many small units rather than a single giant aerogenerator (as has so far generally been the preferred policy in the UK and West Germany) is a more flexible way of meeting sudden fluctuations in demand or shutdowns for the servicing and repair of power stations. Construction lead-times are only 3 months as opposed to 5–10 years for conventional power plant, and maintenance problems are also significantly reduced.[9] The State of California is now far less dependent on single supply sources and enjoys a decrease in its energy expenditure and levels of pollution. During the 1980s, of the 1,660 megawatts of wind-electric capacity which was installed worldwide, 85 per cent was in California, demonstrating again that it is possible to introduce flexibility and diversification of supplies rapidly, once policy objectives have been established.

Japan and Israel are also diversifying their energy supply sources. With its heavy industry and few indigenous fuel reserves, Japan has become one of the world's pioneers in developing new, solar-related energy technologies, including wave power and ocean thermal energy conversion (OTEC).

Finally, the flexible integration of energy supply systems will be particularly vital in the longer term, when the introduction of renewable sources into regional grid systems gains momentum, as this will help to compensate for the seasonal fluctuations in the different supply sources and variations in demand.

7. The Principle of Low Risk

The principle of low risk ensures that energy-policy planners make a thorough assessment and valuation of all conceivable risks and their probable social, economic and environmental consequences before selecting a particular type of energy supply system.

Most conventional systems of fossil fuel extraction, energy generation and even some alternative energy forms carry an element of risk. The obvious risk is failure of supply. Accidents also occur.

Conventional accident evaluation methods must also be revalued from time to time in the light of current data. To the statistical reckoning of *frequency* and *magnitude* must now be added *duration*, a comparatively new environmental factor. Its concern is with the duration of the *effects* of an accident on the community and the social and economic consequences. All these hazards should be assessed against the risks inherent in competing energy technologies.

In recognition of the complicated nature of many modern technologies, the United Nations Environment Programme (UNEP) has produced a handbook which, it is hoped, will assist decision-makers in government and industry to make the most appropriate policy and technological choices and to respond more effectively should accidents occur. The report is entitled APELL (*Awareness and Preparedness for Emergencies at Local Level*). Charles Perrow's book *Normal Accidents—Living with High-Risk Technologies* is more fundamental in tackling the roots of risk.[10] This study, published in 1984, *before* Chernobyl, cogently demonstrates that the more complex the technology, at whatever degree of sophistication, the higher and more incalculable are the dangers. In differentiating between 'incidents' and 'accidents', Perrow stresses that incidents are normally due to failures at component or unit level, while accidents occur through failures of entire systems or even sub-systems. His study documents 12 major, world-wide accidents which, according to normal risk assessments 'could never have happened', including that of the Harrisburg reactor on Three Mile Island.

From current statistical data on different types of energy system one can safely conclude that most renewable systems carry a considerably lower risk element than the fossil fuel options and above all, nuclear power. Within the conventional supply sources themselves, it appears that smaller is generally safer, on the assumption that the basic technology and maintenance regulations are sound. Again, small-scale, locally-based systems have the advantage of locally-based responsibility and avoid some of the 'human element' hazards which increase where technicians are far removed from the decision-taking centre, when a work force's morale becomes low, or when cuts are made in material and safety standards from 'head office' due to organizational and economic constraints.

8. The Principle of Evaluation

Each local community is required to measure the quality and quantity of its energy usage against agreed levels and standards of responsible consumption.

The eighth principle of energy policy recognizes that, for a sustainable energy policy to work, it is vital that there is full, fast and frequent feedback of consumption data against agreed energy-saving targets.

It is my personal conviction that *total* energy accounting should be introduced into every sector. Instead of evaluating goods and services in financial terms, energy accounting measures them in terms of the energy consumed in their production and operation. Understanding the total energy consumed in the process of obtaining raw materials, manufacture, construction, transportation, installation and operation together with the life expectation would help to evaluate the energy viability of a particular installation as against a possibly more energy-efficient alternative. Dr Brian John, for example, points out that 'modern series-produced wind turbines can repay the energy used in a period of 18 months', while 'a nuclear power station consumes more energy than it ever contributes to the national grid' when the energy costs of fuel processing, waste disposal and decommissioning are added.[11] The same is true for certain other forms of generating and co-generating energy schemes.

Finally, in evaluating energy policy, the *social* costs of energy consumption should be added to the economic costs before energy pricing decisions are made. For example, an EEC study by O. Hohmeyer of the social costs of energy consumption in East Germany prior to unification has demonstrated that without including social costs the market is seriously distorted in favour of nuclear electricity and fossil fuels and against the renewable energies. The study asserts, for instance, that when the total costs of electricity generation to society are analysed and evaluated, wind generators were already the most cost-effective of all energy-generating technologies by 1985.[12] The social costs of the 1986 Chernobyl accident, on the other hand, will be economically and socially incalculable even during the lives of our great-grandchildren. We thus appear to have struck a Faustian bargain on their behalf.

Conventional economists should note that, according to conventional economic theory, each rational decision has to be based on an underlying economic function. In the case of individuals, economic theory assumes this objective function to be *utility*, in the case of private enterprises, *profits*, and in the case of political institutions, *social welfare*, all three including present and future aspects. Social welfare is usually understood to be the value of well-being of all individuals in society, including that of future generations.

9. The Principle of Social Justice

The principle of social justice recognizes that affordable access to energy supply is a basic human right.

Justice is concerned with equality of rights and opportunities and the fair treatment of persons or groups in relation to one another. It has probably never been embodied in any energy policy. The right to energy is as important as the right to food, clean water, shelter and clothing. Social justice is, therefore, a prerequisite of any organic energy policy and one should examine what constitutes a just energy policy for a given community in a given situation.

Social justice in energy policy terms would include the rights: of equal access to energy supply; of buying or selling energy at fair prices; of appealing against energy monopolies which distort prices; of avoiding discriminatory tariffs; of information about true costs, environmental impact and other relevant data of various energy technologies; and, finally, the right of protection for a community suffering environmental damage through energy generation elsewhere.

When we examine the principle of social justice in terms of energy policy, we soon see that it rarely exists. In Third World countries the lack of it is most apparent, since only the rich have access to electricity and oil. Fuelwood from the countryside is sold at extortionate rates in the cities, often leaving the surrounding area denuded of trees. Women and children often have to travel many hours on foot to collect enough wood for essential cooking of food and boiling of water.

Even in industrial countries, social injustice is very apparent, since the rich will heat their swimming pools and make unnecessary car and air journeys while the poor often do not have enough money for basic heating and lighting requirements. In another respect, although the US has set an example with its Freedom of Information Act, in most countries it is impossible to obtain accurate information about many types of fuel. Either the information does not exist in an accessible form, or is deemed to be 'commercially sensitive' or, as in the case of the nuclear industry, is deliberately enshrouded in secrecy.

At another level of social injustice, the health, environmental and economic costs borne by many countries in Western Europe as a result of the Chernobyl disaster, and the pollution and environmental degradation of many East European countries, resulting from power generation in Western Europe and aided by the prevailing westerly winds, demonstrate the need for protection against the effects of energy generated outside the community. There are, however, signs of hope. When a local community takes responsibility for its own energy policy, as is the case in Davis, California, the National Centre for Alternative Technology at Machynleth in Wales or even the Cornwall Energy Plan already mentioned, social justice automatically ensues. In this respect all small-scale renewable-energy systems tend to promote a greater degree of social justice. The larger systems, however, such as large hydro-electric schemes, tidal barrages, and off-shore wave power devices, are prone to many of the same inadequacies as conventional sources of energy, although they may be regarded as more sustainable.

10. The Principle of Sustainability and Replenishment

The principle of sustainability recognizes that we cannot continue to live off energy capital but rather that we should live on energy income, so as to meet the essential global energy needs of present and future generations.

This principle embraces all the others since it places them in a long-term, ethically responsible perspective which seeks to conserve for our children and grandchildren. It also recognizes that emphasis must rapidly be shifted from the short-term supply side to the long-term basic demand side of the equation. A sustainable or organic policy is in tune with the self-sustaining and gentle rhythms of nature, and contrasts with the short-term three-to-five-year political and business time scales common to industrial and urban policy-making, and even the medium-term five-to-ten-year lead times acceptable for large-scale technical and engineering projects.

An organic energy future will maximize the use of those natural, solar-derived energy flows, such as direct and passive solar energy, wind power, water, tidal and wave power, possibly geothermal and hydrogen technologies and, above all, biomass, which is the most versatile substitute for oil. All these are continuously replenished. The potential for renewable-energy utilization is vast and far exceeds the world's total energy requirements, even with the future expectations of a growing world population.

Sustainable policies must recognize that fossil fuels are a scarce resource and, until replacements are found, should focus on essential demand 'needs' rather than assumed 'wants'. Most industrial and many Third World societies are, however, dependent on unsustainable, non-renewable energy resources which will not provide for the needs of the next century without ecocide. What is more, non-renewable energy resources cannot possibly cater for the exponential energy needs of an ever-increasing global family.

Lester Brown, Director of the Worldwatch Institute, forcibly pointed out in the 1990 *State of the World Report*[13] that the Earth is reaching the limits of her carrying capacity. Unless there is a concerted re-ordering of Government priorities, in which direct solar conversion, the growing of energy crops on marginal lands and the use of other renewable energy forms, including geothermal energy, are pivotal, there will be no possibility of our biological life-support systems surviving the next 40 years. Edward Goldsmith in *Five Thousand Days*, also published in 1990, is a great deal more pessimistic and implies that we may already be too late.[14] In either case, the time to put the principles into practice in a concerted global strategy is *now*.

Most fossil fuels and uranium reserves are, as we all know, unevenly distributed throughout the world and depend on highly centralized supply systems, capital-intensive investments and sophisticated technical and distribution infrastructures. Thus they are dangerously susceptible to political, social and economic upheavals. Such fuels are in no way able to meet the basic energy needs of most poor communities of the Third World. It is evident, therefore, that the industrial countries' energy policies fly in the face of these 'ten energy commandments' or principles.

We can no longer plead ignorance. From both historic and dramatically recent perspectives, it is plain to everyone that the current industrial societies' demand 'norm' is inequitable and unsustainable in terms of energy supply.

An Organic Energy Strategy for the Future

The emphasis on future energy policies must be to develop the renewable energy forms as quickly as possible, as well as to implement and support all the efficiency and conservation technologies already in existence. Looking at the synergies, we see that, while biomass can not only provide energy but also absorb CO_2, solar energy will be more effective in a warmer climate created by the greenhouse effect, but will not contribute to further greenhouse emissions.

It is interesting to note that Denmark has already embarked on a policy of sustainability and plans to produce 10 per cent of its total electricity requirements from wind by 2000. The world's first offshore wind-energy farm is currently under construction off the coast of Denmark and could possibly be used as a model elsewhere.

As mentioned previously, biomass is not only the most flexible of all the renewable energy forms in the wide variety of its end uses, but is also the most necessary, since plants and trees lock up CO_2 through photosynthesis. This is released back into the atmosphere only when the fuel, in whatever form, is burned. However, no more CO_2 is released than that which was originally absorbed. It is, therefore, necessary, among other criteria, to plant *more* trees than those required for energy purposes to maintain a positive CO_2 balance. The American scientist Gregg Marland has estimated that seven million square kilometres of forest could absorb all current emissions of CO_2 from the burning of fossil fuels, whilst Norman Myers, the British environmentalist, suggests that 3 million square kilometres might suffice, since the phytoplankton of the oceans are believed to absorb up to half the emissions. It is, of course, extremely important to select the type of plants and trees most suitable for a given location and to avoid planting monocultures. It is most important to avoid the dilemma of land use for fuel or food, and this seems to have posed an insurmountable difficulty in the past for policy makers, agronomists and those in charge of forestation projects.

The most positive biomass developments for the future appear to be in agroforestry schemes, which will provide both food and fuel, fertilizer through crop residues, and soil and water protection, as well as intensive local employment and involvement. The importance of agroforestry on a global scale cannot be sufficiently emphasized in an organic energy policy. Herbert Girardet has pointed to several poor Third World communities which have recently achieved self-sufficiency in both food and fuel, such as the Chagga people in Tanzania and the villagers of Uraim in the Brazilian Amazon.[15]

A particular example of how the use of biomass also fulfils the principle of sustainability and replenishment is the recent decision of the US company Applied Energy Services which, to compensate for increased CO_2 emissions that will emanate from a new coal-fired power station being built in Connecticut, is to embark on a programme to plant 52 million trees in Guatemala. This biomass crop will, during its lifetime, absorb an equal amount of CO_2 from the atmosphere and thereby maintain the CO_2 balance. Such replenishment schemes could act as models for sustainable energy policies during the period of transition from fossil fuels to renewable energy schemes.

Conclusion

Principles such as the 10 'energy commandments' can provide a framework within which both to judge and formulate energy and other policies in all main areas. These principles are complementary and convergent and could lead to a more integrated, sustainable and organic future.

Policy makers, energy suppliers and energy users in the past have failed to recognize the synergies which affect all human and societal existence. Long-term integrated regional and national energy policies based on the 10 'energy commandments' will have far-reaching environmental consequences globally. Taken together, the 10 principles will allow an evolutionary movement from the excessive energy dependence of today to an energy-conserving and responsible self-reliance tomorrow.

Organic energy supply systems, as with all else, stem from a global recognition of the interdependencies and interlocking between all sections of human activity, including energy, economics and environmental issues, supported by the need to educate, encourage and set examples. Policy in these other areas must therefore be based on the same variety of principles as govern energy policy so that full synergy is obtained.

The urgent requirement of an organic energy policy is, therefore, to increase Third World energy availability on a sustainable and self-sufficient basis, whilst ensuring that industrialized countries find ways to reduce demand without impairing standards of living. There must be a rapid switch to renewable-energy systems. The technology is there. Much valuable research has been done; what is needed is the political will to implement it. Above all, there must be a concerted effort by Governments, institutions and individuals to reduce waste in every section of human activity, whether in terms of energy or raw materials. Where 'waste' materials are part of the natural order they should be recycled or used as a resource. Neither the planet nor the poor can sustain any more of the effluence of affluence.

Sound principles lead inevitably to good practice and it may be expected that many seemingly intractable and interconnected problems will wither away simultaneously. By contrast, 'carpe-diem' energy policies based on short-term economic benefit and relying on harsh, 'hurricane' technologies are unsustainable and must be abandoned in the light of the global dysfunctions they create. At every level, choices have to be made, and in some instances it may mean choosing between the lesser of two evils. However, energy is primarily about people and not about economics. Humanity's *real needs* must determine the preferred choice. Unreal money values, as in times of war and ecological crisis, will change overnight. The choice is ours now.

Chapter Seventeen

The Healing of the Earth

Richard Grantham

AFTER BILLIONS OF YEARS of geophysical and biological determinism, global changes now have an anthropogenic component. Industry and human numbers have reached a magnitude such as to affect the composition of the atmosphere, the sea level and the mean temperature. The source of this action is a kind of chemical pollution. Thus, modern air contains more and more Earth-warming and climate-perturbing trace gases that for tens of thousands of years had much lower concentrations. This accelerating build-up is called the Greenhouse Drift[1]. These gases may be normal constituents of the atmosphere, such as carbon dioxide or methane, but which have grown in amount in the last two centuries, or they may be synthetic molecules that were completely absent until this century such as the chlorofluorocarbons (CFCs), carbon tetrachloride and bromine derivatives[2]. These bromine compounds are powerful halocarbons, in the same class as the CFCs in ozone-destroying capacity. Envisaging the overall chemical functioning of the planet as analagous to the physiology of an organism, we can conceive of the present global pollution as representing a threat to the health of the biosphere. It is, therefore, appropriate to consider geotherapy.

No rapid cure is possible, however. The Earth system is too vast and complex. Nevertheless, the polluting process is accelerating and some of the trace gases (carbon dioxide, halocarbons, methane and tropospheric ozone) already have alarming rates of growth. These gases can damage the health of humans, animals and plants as well as adding to the greenhouse effect. Likewise, our groundwater, lakes and rivers, and even the ocean, are becoming poisoned by acid rain and excessive use of fertilizers and chemicals. Hence, we cannot dilly-dally. Most scientists believe that immediate action should be taken to reduce CO_2 emissions, despite the scientific uncertainties that remain. Industry and our lifestyle have unconsciously been disturbing the composition of the air in the last two centuries. As foreseen by several scientists, especially the brilliant Swedish chemist Svante Arrhenius,[3,4] perturbing the atmospheric amount of water vapour, carbon dioxide or other trace gases causes changes of ground temperature and climate. Suggested remedial measures that would lead to better health of the global ecosystem follow.

First, it is worthwhile to distinguish adaptation and correction. The former is seen by some global-change scientists as the only possible posture before the increasing greenhouse drift. That is, people would move to higher ground as sea level rises, sea walls would be built, houses would be better insulated and air-conditioned, and so on. The second approach demands a direct attack on the causes of the drift: regulating global photosynthesis to bring fixation and consumption of carbon into balance with emissions. In fact, this correction implies extending and caring for the primary production of all natural ecosystems (forests, savannahs, coral reefs, mangroves, soils, etc.). Both kinds of treatment are necessary; adaptation to lower carbon-consuming energy use in all industrialized countries is desirable and, at the same time, chemical and biological pollution can be significantly reduced in most regions. Preventive geotherapy would operate on a longer time scale and require planning for the future, since part of its task should be to correct errors of the past. For example, in modern times a huge amount of carbon stored as fossil fuel for hundreds of millions of years has been converted to CO_2 and now resides in the atmosphere or ocean, far surpassing the capacity of natural ecosystems to photosynthesize it. This excess CO_2 may need to be partly re-converted to organic carbon to avoid future global heating, since all planetary-scale processes are characterized by a variable, and ordinarily unknown, lag time. The Earth has warmed by about 0.5C° so far this century, but if we were to limit CO_2 emissions so as to maintain the present content in the air of 350 ppmv (parts per million by volume), we would have no guarantee that additional warming would not take place. Therefore, establishing a global balance between photosynthesis and oxidation (combustion plus respiration) of carbon is urgent to avoid further warming. We should also prepare to lower the current level of all trace gases in order to minimize their continued contribution to the greenhouse drift in the next few decades.[5]

A conscious evolutionary strategy

The most important long-term measure, which could serve as a framework for all interventions in nature, would be the adoption by society of a conscious evolutionary strategy to replace our present will-less and direction-less determinism. It is not that we ought to regiment human activities or engineer the biosphere, but it is possible to raise our chances of survival by agreeing on geotherapy. Indeed, it is necessary that certain processes be brought under control. With the present extent and state of our civilization and human numbers, and their rate of increase, the processes of greenhouse drift, stratospheric ozone layer depletion and pollution in general are already threatening the physical, chemical and biological well-being of the human species. In 100–200 years, if the greenhouse drift is not controlled, the very existence of higher organisms may be put in question.

The question then is, how do we achieve this control? We evidently need to go beyond national politics, of which there are hundreds of varieties, and some sort of global mechanism is required. The United Nations could be a covering structure, but it has not provided a universally agreed philosophy for planetary-scale action.[6] Human society has had examples of environmental control and remedial action for more than two centuries,[7] but these have always been national enterprises. Nevertheless, the rationale for global correction can be expressed in terms of the human future. How can humans continue to inhabit this planet? It is clear that several actions must be taken to ensure this continuity. They are, however, *global actions*, requiring a global reason for their doing, and all peoples should be in agreement on their goals and manner of execution. The only way to accomplish them is to surpass present politics and power structures. The biosphere is a scientific phenomenon. We can, therefore, base decisions on biological concerns and the evolutionary process, which, as scientific matters, are less subject to contention. When there is discord, a peaceful method is available for settling differences. The science and philosophy of the evolutionary process are not linked to the economic or political status of a nation or group of nations, and fortunately no military-based extension of scientific policy exists.

This does not mean giving the greenhouse problem over exclusively to evolutionists. What I am suggesting is that *evolutionary thought* is a desirable element in planning the future and that our chances of surviving will be greater if we plan. A conscious evolutionary strategy would give humans a goal: to better conditions in the biosphere, thus assuring its continuity. World-wide education is, of course, needed to prepare the change.

With this idea as a general framework for the task of greenhouse correction, let us consider the main specific problems and propose some solutions. The six main greenhouse gases (water vapour, carbon dioxide, methane, the halocarbons, nitrous oxide and tropospheric ozone) and other interacting airborne molecules (carbon monoxide, oxides of nitrogen and sulphur, and peroxides[8,9]) all require regulation. To start with, the desired global concentration range for each of them should be agreed on. In addition, the hydrological cycle needs more attention, particularly since we do not know whether relative humidity will rise or drop as trace gas amounts and temperature increase. The way it changes is critical to precipitation, runoff, erosion and desertification.[10] For each of these problems concrete steps should be taken without delay.

Steps in correction

Nitrous oxide (N_2O). Build-up of this gas is linked to growing nitrate fertilizer application and forest felling.[11] Even in areas where population size is stabilized fertilizer use is accelerating: for example, in Europe it has doubled in the last 25 years; in other regions the increase has been greater, having gone up by a factor of 10 in Asia.[12] Nitrous oxide is also produced in combustion, either of biomass, such as trees or grass, or of fossil fuels, as in internal combustion engines. Many natural habitats release it in widely varying amounts but some soils may be a sink for N_2O—that is, they may absorb it,[13] although little work has been done on this aspect. Whether denitrification is the main source of oceanic nitrous oxide remains controversial.[14] The clearest mandate here is for reducing chemical fertilization, especially its *excess*, which leads to runoff contamination and poisoning of lakes, rivers, groundwater and ultimately the ocean.

Nitrous oxide deserves more measurement in various habitats and processes; its long atmospheric residence time (160 years) and global-warming potential per molecule, 180 times that of carbon dioxide, make it a worrisome gas.[15]

Methane (CH_4). The smallest among important infra-red absorbing molecules, methane may be the worst to control because of its diverse sources. Recommendations for controlling it have already been given.[16,17,18] In brief, the main sources are paddy rice growing and ruminant digestion. Both are increasing rapidly and should be curbed, since they can be managed more easily than natural wetlands, which also release much methane.[19] Emissions from mines and waste sites likewise need to be limited.

Although the atmospheric mass of carbon in methane is only 1/200 that of carbon dioxide, airborne methane is growing much faster than carbon dioxide. During the last glacial period, which lasted over 100,000 years, its mean concentration was 400–450 ppbv (parts per 10^9 by volume) and the lowest values (350 ppbv) occurred at the Last Glacial Maximum, around 20,000 years ago. At the interglacial peaks, of roughly 130,000 and 10,000 years ago, it reached about 700 ppbv. In the last three centuries, however, the concentration has shot up from 700 to 1700 ppbv, nearly 2½ times the highest value in the total Vostok ice-core record of 160,000 years.[20]

Per molecule, methane absorbs 20–30 times more infra-red energy than carbon dioxide, but fortunately it has a much shorter lifetime than CO_2 (14 and 230 years, respectively),[21] otherwise correction would be more difficult. Methane's lifetime depends on its conversion to water and CO_2 by the hydroxyl radical (OH), which is made from water by photolysis. Many molecules, including water, are broken down in the upper stratosphere by high-energy solar rays in the process called photolysis. Hence, future water and methane concentrations appear critical, especially since atmospheric carbon monoxide (also an OH sink) is increasing as well.[22] It is important, therefore, that the methane risk be lowered by limiting emissions and exploiting natural sinks for it, such as silviculture on sandy soils[23] and biological oxidation in tundra.[24] More water in the stratosphere might seem desirable to assure sufficient OH for keeping methane and carbon monoxide levels down, but increasing water there is dangerous, because it adds to the greenhouse drift and it forms ice clouds, which can serve as substrates for ozone destruction.[25]

Finally, burning natural gas, which is mainly methane, as a substitute for petroleum and coal, would have two short-term advantages. First, its combustion gases contain much less sulphur than desulphurized flue-gas from coal burning and would thus add less to acid rain. Second, a molecule of methane is 25 per cent hydrogen by weight, hence the amount of CO_2 produced per unit of energy by its burning is less than with petroleum and coal, which have higher C-to-H ratios.[26, 27] With time, however, natural gas is exhaustible and its use should be limited when solar, eolian, hydrogen and other sustainable energies can replace fossil fuels.[28]

Ozone (O_3). Human activities are increasing the amount of ozone, a strong greenhouse gas, in the lower atmosphere. At mid to high northern latitudes, ground-level ozone has apparently doubled in the last century and its rate of increase is expected to grow in the next 30 years. This acceleration is anticipated because of increasing emissions of nitrogen oxides and methane, which stimulate tropospheric ozone formation, especially at 30–90°N, where its production has quadrupled during the last 100 years.[29] Controlling fossil fuel burning is required to lessen production of ozone precursors.[30, 31] In contrast, stratospheric ozone, the biosphere's ultraviolet filter, is vanishing, hence menacing our food supply and the health and genetic stability of humans, plants and animals. Thus, two ozone problems exist, with different causes and locations; both relate to global heating.[32]

Halocarbons. The CFCs and other halocarbons are active throughout the atmosphere as greenhouse agents and in the stratosphere as ozone destroyers. CFCs are eventually removed by photolysis in the upper stratosphere, but several have lifetimes of more than a century; replacements with shorter lifetimes are being evaluated.[33, 34] The relative contributions of CFCs and CO_2 to total global warming during the 1980s were reported to be 25 per cent and 57 per cent, respectively.[35] More recent work has revealed, however, that CFCs can have either a net warming or cooling effect, varying with altitude; if so the US policy of depending on reduction of CFC emissions to lower total greenhouse drift from all gases would not be credible.[36, 37]

It has also been shown that carbon tetrachloride and methyl

chloroform are important anthropogenic greenhouse gases.[38] All halocarbons, whether chlorine or bromine derivatives ('halons'), can contribute to Earth warming and ozone layer destruction. Their production and use, along with that of the CFC substitutes, can be regulated by the updating of, and world-wide adherence to, the Montreal Protocol. (The Montreal Protocol is the starting agreement among nations for controlling substances that deplete stratospheric ozone. The meeting was held in September 1987; so far several dozen countries have signed the 'treaty'. It is a landmark in international co-operation and can serve as a model for future control of other gases. It first concerned just the CFCs, but recently it has been updated to include carbon tetrachloride, methyl chloroform, etc. It needs to be extended and enforced.)

Carbon dioxide (CO_2). The CO_2 content of the air has increased by 20 per cent in the last century.[39] Atmospheric CO_2 load and global temperature co-vary; at least over the last 30 years these parameters are significantly correlated.[40] Therefore, along with the Vostok ice-core data,[41] we have both short-time and long-time evidence of the correlation. The greatest help for the flagging carbon cycle would be to build up photosynthetic capacity by caring for and extending all natural ecosystems. Large scale afforestation on all continents, especially Africa, is necessary for establishing a balance between consumption and fixation of carbon.[42, 43, 44, 45, 46] It is not sufficient, however, and the other major ecosystems should be included in the global treatment.

Water (H_2O). Water vapour accounts for most of the greenhouse effect and it has been associated with positive feedbacks that can amplify Earth warming.[47, 48, 49] Soil wetness is tied to relative humidity and, less evidently, to CO_2 concentration. As the CO_2 amount and temperature rise, the soil will become drier over large parts of the Earth[50]. This would constrain any fertilizing effect from higher CO_2 amounts and could be disastrous in some regions. Water sequestering can be improved by continental management including groundwater replenishing, biomass increase by afforestation, stopping over-grazing and grassland burning, controlling runoff and erosion, and caring for mangroves and corals, all of which would promote photosynthesis and help reclaim deserts. A long-term effect on evapotranspiration, hence relative humidity, can be sought by controlling the extent of the Earth's surface covered by trees and other vegetation. The ecological impact of these large-scale projects must be studied with care.

Consequences of inadequate action

If strong action is not taken, world industry, population and pollution will continue to grow.

The greenhouse drift will accelerate owing to the faster build-up, relative to CO_2, of methane, halocarbons, tropospheric ozone and other trace gases, bringing increases in global temperature, sea level and storm extent and intensity. These predictions are still based mainly on model studies; much uncertainty remains on details, but the general pattern is a matter of scientific consensus.

Soil degradation and desertification will increase, constraining agriculture and demanding more irrigation and fertilization, both of which can degrade soils and promote pollution, worsening this vicious circle.

Developing countries will add greatly to the greenhouse drift, water shortage, and global pollution; in the short term their energy will have to come from biomass and fossil fuels, since nuclearizing their electricity industry, if technically feasible, is illusory within the next 50 years, owing to cost, weapons risk and waste disposal problems. In addition, sustainable alternatives such as wind, sea or solar power are being neglected in these countries.

Under the above conditions, it is hard to believe that the present period of relative peace can last, even though the superpowers disarm and try to banish war. Economic and ideological violence may grow.

Species diversity will decrease greatly and our own survival appears problematical. The palaeontological record shows that major climate change is always associated with a major redistribution and extinction of species. There is no way of predicting which species will benefit, survive or perish in such events; frequently the dominant species is replaced.

Proposition: A World Ecological Council

Several courses are open to us:

(*a*) Have faith that some extraterrestrial agent will save us;

(*b*) Do nothing, allow nature to take its course;
(*c*) Prepare to die out gracefully as an unadapted species;
(*d*) Begin national or local palliative and curative measures;
(*e*) Institute scientifically-coordinated world correction.

Only the last possibility is believed capable of leading to balance and harmony for the whole planet. Our task, then, is to manage the correction, the first step being to set up a structure and mechanism for effecting global correction.

The proposed World Ecological Council (WEC) should not be a body for promoting procrastinating research instead of taking needed action. Its essential mission would be the development of a strategy for the future of humans and the biosphere, and the coordination of all corrective measures.

A Commission for the formation of the World Ecological Council is suggested. For the WEC, the Commission would: define the objectives, responsibility, function and structure; suggest the number and composition of the members; recommend a budget and source of financing; and write a report responding to the three above points and indicating desired initial actions of the WEC, as well as any other matters judged appropriate. The report would be sent to the United Nations, with copies to all heads of state. Members of the Commission would be chosen by continent, with each continent contributing six members. The continents would be Africa, Asia, Europe, North America, South America and a sixth 'continent' consisting of Australia, Indonesia, Japan, New Zealand and Oceania. Each continent would decide on its own system of nominating its six members, which should be completed within one year. In case of conflict over nomination, the United Nations would be charged with selecting the member. The Commission would thus have 36 members, from which a president would be elected by vote of the members. The Commission members should possess a high and broad degree of scientific literacy but need not all be scientists, since the general public and the disciplines of macro-economy and philosophy should be represented.

Concluding remarks

How to promote, effect and render profitable a global vision is the question of our time. The great danger in globalization of our resources is that no one will have the responsibility for them and they will all be over-exploited.[51] Furthermore, future generations deserve consideration. The theory of inter-generational equity says that present humans hold the natural and cultural environment of the Earth in common, both with other members of the present generation and with other generations, past and future.[52] It seems to me that these ideas are compatible with those I have exposed and that the appropriate global framework for their realization should be sought.

I end with six principles that colleagues and I believe necessary for successful correction of the greenhouse drift. This list is our approach to gaining agreement among all peoples on what we desire for the corrected planet and what attitude we can have in changing the present one.

Greenhouse drift correction principles

1. Seek the long-term health and stability of the biosphere by caring for and extending natural ecosystems. At the same time, existing natural resources should be conserved and when possible replenished.

2. Act to improve the environment generally, as well as to correct the greenhouse drift. For example, reducing tropospheric ozone pollution would improve living conditions *and* lessen Earth warming.

3. Promote a balance of indigenous species in all actions. Such species have proven their adaptation in specific regions, many of them over millions of years. Re-installing them in their own territory risks less than introducing foreign species.

4. Encourage biological diversity, avoiding monoculture. Thus, in managing degraded areas, as in afforestation of deserts, mixed (indigenous) species are indicated.

5. Bring human numbers to a sustainable level, that is, within the long-term carrying capacity of the Earth.

6. Institute a framework to apportion the many tasks. The willingness

of different individuals and groups to participate in the programme will depend on a perceived fairness in the distribution of burdens and benefits. Let us accept a moral imperative to take care of global ecosystems by just and impartial means. Otherwise co-operation will inevitably break down.

To heal the Earth we will need wisdom as well as knowledge. The Geotherapy group of the Global Change Project of the International Union for Quaternary Research has begun to work more directly on the concept of a Global Bioethics.[53] We recognize the need for a cross-cultural philosophical system to make geotherapeutic measures understood by and acceptable to all peoples.[54, 55]

*

I thank H. Faure, T.J. Goreau, T. Greenland and E.G. Matthews for collaborating on the manuscript.

Chapter Eighteen

The Organic in Theory & Practice

Philip Conford

BECAUSE THE ORGANIC PHILOSOPHY sees things as a whole, it opposes mechanistic and reductionist attitudes to nature. It takes the view that we do not understand a whole by breaking it down into its elements, but that, on the contrary, we understand the part only when we see its relationship to other parts and to the whole. It emphasizes synthesis, being concerned with the way in which different parts and processes are integrated. As Jeremy Naydler expresses it, 'contexts are seen as prior to causes, and the interrelationships between phenomena are grasped as dynamic wholes rather than analysed into their component parts'.[1] This division between mechanistic and organic interpretations of the world can be traced back to classical Greece. 'Ever since Democritus sought understanding in *atoms*, and Plato and Aristotle in *forms*, there has been a vigorous competition between two sets of ideas: atomism—material analysis—quantitative precision, and form—unity—symmetry...'[2] The biological chemist L.J. Henderson described Aristotle's 'conception of the living thing as an autonomous unit in which every part is functionally related to every other and exists as the servant of the whole'.[3]

Advances in mathematical and physical science during the 17th century gave great impetus to the idea that both nature and humanity could best be understood in mechanistic terms. The 'root metaphor', to use Donald Worster's phrase, of the world as a machine, lent authority by Newtonian physics and the philosophy of Descartes, dominated European thought.[4] One of this symposium's contributors, the Indian scientist Vandana Shiva, contends that its effect has been to create a 'world-view in which nature is: (*a*) inert and passive;

(*b*) uniform and mechanistic; (*c*) separable and fragmented within itself; (*d*) separate from man; and (*e*) inferior, to be dominated and exploited by man'.[5]

The metaphor was powerfully challenged, around the end of the 18th century and early in the 19th, by the Romantic movement, for two main reasons. As the biological sciences advanced, the conception of the world as a machine was deemed, by a number of influential thinkers, inadequate to explain the growth and development of living organisms. The other reason was a sense that conceiving of nature as inert and passive robbed it of its beauty and its power to inspire reverence and love. The practical application of the mechanistic outlook resulted in the ugliness and lifeless uniformity of industrialism, which were to be attacked by Dickens in *Hard Times*.

Donald Worster describes Romanticism as 'fundamentally biocentric'.[6] The defining features of organic form in Romantic thought have been identified as: the priority of the whole; the manifestation of growth; the conversion of diverse materials into the form's substance; the self-determination of the form through an organism's own energy; interdependence of parts; and 'the tension and reconciliation of manifold opposites...'.[7] Goethe, an early investigator of environmental influences on plants, undertook 'the search for a method of perception that would represent nature "at work and alive, manifesting herself in her wholeness in every single part of her being"...'.[8] In England, Coleridge's philosophy, deeply influenced by the German Romantics, expressed the same idea. 'Coleridge's central preoccupation was with the antithesis between a living whole or organism on the one hand and a mechanical juxtaposition of parts on the other.'[9]

A century later, the battle was continuing, as Vitalists engaged with Mechanists and produced major works of biological philosophy. One of the most influential was *Holism and Evolution*, by the South African statesman J.C. Smuts, which expressed the belief that 'The creation of wholes, and ever more highly organized wholes, and of wholeness generally as characteristic of existence, is an inherent character of the universe'.[10] Concerning the nature of an organism he wrote: 'An organism... is a natural whole... It consists of parts, but its parts are not merely put together. Their togetherness is not mechanical... if these parts are taken to pieces the organism is destroyed and cannot be reconstituted by again putting together the severed parts. These parts are in active relations to each other...'[11] In the same decade, the 1920s, the philosopher N.O. Lossky developed a similar metaphysic. 'Those who take an organic view of the world conceive of the whole universe after this fashion: they regard every element of the world... [as] existing, not independently, but only on the basis of a world-whole, only within a universal system. All our usual ideas about the world, suggested by the atomistic and mechanical way of looking at it, are then reversed...'[12]

This central idea of the organic outlook, the primacy of the whole, has been an important influence in various areas of 20th-century thought: for example, in ecology, social and political theory, systems theory, and *gestalt* psychology. Lancelot Law Whyte, in the 1968 edition of *Aspects of Form*, believed he saw a 'new unifying discipline' emerging, whose *leitmotif* could be summarized as the view that 'This is a world of form and structure and can only be properly understood as such'.[13] The scientist Rupert Sheldrake has written more recently of 'habitual organising principles [which] may be inherent in all nature'.[14] For Ronald Harvey, the hologram is an image of this truth, since 'In the hologram every part of the whole contains or implies the whole'.[15]

The primacy of the whole necessitates the interconnectedness of the parts. Wendell Berry describes the healing process as a matter of 'restoring connections among the various parts... When all the parts of the body are working together... we say that it is whole; it is healthy.'[16] His fellow American and fellow farmer, Wes Jackson, sees 'a profound awareness of the total interpenetration of parts'[17] as the essence of the ecological world-view, and Vandana Shiva expresses the same idea: 'Modern reductionist science... has... excluded ecological and holistic ways of knowing which understand and respect nature's processes and interconnectedness *as science*'.[18]

As we would expect, the importance of interrelatedness is to be found in Romanticism. Worster summarizes the Romantic approach to nature as 'concerned with relation, interdependence and holism';[19] an attitude exemplified by Wordsworth's condemnation, in *The Excursion*, of the reductionist mistake of 'Viewing all objects unremittingly / In disconnection dead and spiritless' (Book IV, lines 961–962). 'The Romantics saw nature as a system of necessary relationships that cannot be disturbed in even the most inconspicuous way without changing, perhaps destroying, the equilibrium of the whole.'[20] And there is, in this Romantic view, another aspect of interdependence which is of great importance—that of the relation between the known object and the knowing mind. Man and Nature are linked, not detached from each other. Worster speaks of the Romantics' 'desire to restore man to a place of intimate intercourse with the vast organism that constitutes the earth'.[21] All creatures are part of the same living world, and completeness of knowledge comes not as a result of detachment from it, but only with a loving relationship to it.

This vision of interconnectedness was not the preserve of poets. The naturalist Thoreau studied the interrelations between plants, animals, and their environment, and Ernst Haeckel—generally regarded as the founder of scientific ecology—'suggested that the living organisms of the earth constitute a single economic unit resembling a household or family dwelling intimately together...'.[22] Towards the end of the 19th century, the Danish ecologist Eugenius Warming described ecology as being 'concerned chiefly [with] "the manifold and complex relations subsisting between the plants and animals that form one community"... So linked and interwoven... are the lives of all members in the community that "change at one point may bring in its wake far-reaching changes at other points"...'[23]

The results of man-made changes in the United States were seen in the 1930s, when the Dust Bowl was, for the ecologist Frederic Clements, proof that 'man's actions in one place can ramify destructively through a whole biota... [and] across an entire nation'.[24] Rachel Carson was to draw a similar conclusion in the 1960s from the effects of pesticides. Clements expressed scientifically the oft-quoted idea of Chief Seattle, who had foreseen in 1852 the dangers looming when the Indians relinquished their lands: 'All things are connected. Whatever befalls the earth befalls the sons of the earth. Man did not weave the web of life, he is merely a strand in it'.[25]

We shall return in a while to the social and religious implications of the organic outlook, but first we need to mention an aspect of this philosophy that is closely associated with the principle of interconnectedness: the importance of variety and aesthetic harmony. The concept of the organic involves the idea of a unity of differences.

As John Macmurray explains: 'The matter of the living body... is in a continual process of dissolution and replacement. The form is preserved... but with variations... we find that [the organism] consists of a set of parts or elements which differ from one another. These differences are so arranged that they preserve a form... We must represent the unity of what is alive as a unity of differences.'[26]

The philosopher's view is confirmed by the biologist, C.H. Waddington, when he writes that organic form is 'something which is produced by the interaction of numerous forces which are balanced against one another...';[27] and, where cultivation is concerned, diversity is described by Patrick Rivers as 'a kind of insurance policy against loss of any one species through over-swift or excessive change', because 'the greater the number of different species there are in an ecosystem, the more it is able to withstand damage...'.[28]

An organic system, then, is a multiplicity within a unity. If this is so, it is not surprising that, in the rather dismissive phrasing of Donald P. Hopkins, 'the Nature-knows-best philosophy' of the organic school of farming 'tends to appeal to those whose leanings are somewhat artistic...'.[29] As Macmurray says, a blend of unity and diversity 'can be represented only in aesthetic terms, as a balance or harmony'.[30] The language of aesthetics can be almost indistinguishable from the language used in definitions of the organic. Here, for example, is Aristotle talking about the plot of a tragedy: 'its various incidents must be so arranged that if any one of them is differently placed or taken away the effect of wholeness will be seriously disrupted'.[31] In the Middle Ages, 'beauty was thought of as a combination of unity and variety as when diverse and varied parts are combined into a unified whole so that the congruence of the parts in the whole is immediately perceived'.[32]

The 20th-century philosopher, John Hospers, has suggested that organic unity is the major criterion of judgement for the formal elements of a work of art. 'The unified object should contain within itself a large number of diverse elements, each of which contributes to the total integration of the unified whole...'[33] Similarly, the art critic, Herbert Read, argued that patterns in nature are 'significant by virtue of an organisation of their parts which can only be characterized as *aesthetic*...'.[34] He drew the conclusion: 'Aesthetics is no longer an isolated science of beauty; science can no longer neglect aesthetic factors'.[35]

This idea had, of course, already been articulated by the Romantics. For Goethe, botany and aesthetics were linked through their concern with form, and the philosopher Schelling saw the organic as the concept which unites nature with art.[36] There would be no conflict between science and beauty if the organic mode of perception replaced the mechanistic, since the new outlook would reveal the beauty which was to be found in the living world of nature. 'That beauty is... rooted in life is indicated by the fact that... the physical structures of organisms... almost always appear attractive to us and often very beautiful', says E.W. Sinnott;[37] a faith to which support is lent by Ernst Haeckel's collection of drawings of organisms entitled *Art Forms in Nature*.[38] Herbert Read went as far as to state categorically: 'Art differs from nature, not in its organic form, but in its human origins...'.[39] An idea similar in its implications was expressed by the Catholic philosopher E.I. Watkin: 'Beauty has been aptly denominated *splendor formae*—the resplendence of form or pattern'.[40]

But this should not be interpreted as a static concept. As we have already seen, the category of the organic applies to what is living and growing and the ideas of balance, rhythm and dynamic equilibrium are as important to it as ideas of order and pattern. To quote Macmurray again: 'The sense of organic well-being tends to express itself in rhythmic movement... It is certainly the case that there is a close and constant relation between art and the organic aspect of our own experience.'[41] The architect, Frank Lloyd Wright, outlining his theory of an 'organic architecture', based his ideas for the structure of buildings on 'the occult symmetry of grace and rhythm affirming the ease, grace and naturalness of natural life'.[42]

If beauty, rhythm, harmony and pattern are truly features of the world around us, then it is not hard to see why the organic outlook tends to be mystical or religious. According to Macmurray, mysticism is the attempt to experience the whole of reality as an object of aesthetic contemplation;[43] and religion, by definiton, is that which binds together. If one believes in the existence of an order and harmony which are not man-made, reverence towards them will be an appropriate response. To disrupt the pattern becomes an act of *hubris*, an arrogant defiance of the divine order, exemplified in Romantic writing by the Sorcerer's Apprentice of Goethe's tale, and Coleridge's Ancient Mariner.

Belief in some sort of divine order in nature was expressed by Wordsworth in the poem generally known as *Tintern Abbey*, when he wrote of

> A motion and a spirit, that impels
> All thinking things, all objects of all thought,
> And rolls through all things.

In Schelling's philosophy, 'nature... is nothing other than a development towards spirit... spiritual form does not conflict with organic form but is rather the fulfilment, the maturest fruit of the organic process itself'.[44]

In the same tradition, a century later, Lossky wrote: 'A consistent working out of an organic view of the world leads to the recognition of a super-organic principle'[45]—a kingdom of harmony or kingdom of the spirit.[46] Smuts, though unwilling to commit himself on the question of the existence of God, uses unmistakably religious language in describing the nature of his holistic universe: 'The holistic nisus which rises like a living fountain from the very depths of the universe is the guarantee that... the ideals of Well-being, of Truth, Beauty and Goodness are firmly grounded in the nature of things, and will not eventually be endangered or lost'.[47]

Among contemporary writers, Wendell Berry has expressed his 'faith... that things connect—that we are wholly dependent on a pattern, an all-inclusive form, that we partly understand'.[48] Wes Jackson declares himself at ease with the 'Biblical injunction that we are to "dress and keep" the earth... We have one job to do, single-mindedly—to care for the *Creation* that the ancient God of the Hebrews and our own God found to be good and loved.'[49] Vandana Shiva tells us that 'In Indian cosmology... person and nature... are a duality in unity... There is no separation here... between the sacred and secular traditions';[50] an idea comparable in certain respects to Philip Sherrard's theology of nature as a sacrament, filled with God's presence.[51]

Some form of spiritual belief is integral to the organic outlook,[52] and the best-known of the scientific theories associated with the movement—the Gaia hypothesis—has, for many of its supporters, a strong mystical appeal. James Lovelock is explicit that his hypothesis 'is a religious as well as a scientific concept'.[53] 'I have tried to show that God and Gaia, theology and science, even physics and biology are not separate but a single way of thought.'[54] The Gaia hypothesis sees the whole of Earth as an organism—alive, self-balancing, and to be revered. It was anticipated in the 19th century by Theodor Fechner and Wilhelm Preyer, whose ideas were criticized, interestingly enough, by Haeckel. Haeckel said of the latter that 'he somewhat mystically connects the idea of a conscious God with that of a living universe... This extension of the word "organism" has very properly met with little approval in biology.'[55]

Nevertheless, the Gaia hypothesis can be seen as an entirely logical development of the organic starting-point. In A.N. Whitehead's philosophy of organism, nature is seen 'as composed of organisms at all levels of complexity, each one being more than the sum of the parts, and each of the parts in turn being an organism made up of further parts... The earth is a whole system, made up of ecosystems and other sub-systems within it, and the earth itself is part of the solar system... At each level there are wholes made up of parts.'[56] Similarly, in the Gaian philosophy, everything links with everything else and pattern, beauty and equilibrium are fundamental to the universe.

There is one final aspect of the organic outlook—and not the least controversial—to be considered, and that is what it has to say about social and political issues. Clearly, the ways in which human beings organize themselves and the values which underlie their societies have a substantial impact on the natural world. Consumer societies, for instance, will have a particularly disruptive effect on balanced, sustainable approaches to natural resources.[57] As we saw in the Introduction, the British organic movement has never been concerned solely with questions of agriculture and forestry, since these are inseparable from government policies and social changes. Members of the Kinship in Husbandry, such as Lord Lymington, Rolf Gardiner and H.J. Massingham, wanted to see agriculture restored to importance in the nation's economy and social fabric,[58] and can, in a number of respects, be seen as heirs of the Romantics, contrasting rural virtues with the decadent urbanism of a mass society.

The organic husbandry movement was also influenced in its social thought by theories of health: in particular by the writings of Dr Kenneth Barlow, Sir Robert McCarrison, Dr L.J. Picton, and Dr G.T. Wrench, and by the work of the Peckham Health Centre in London.[59] In her study of the Centre, Alison Stallibrass says of its guiding spirit:

'It was part of Scott Williamson's hypothesis that within the ultimate all-encompassing organic whole there are many lesser wholes... of which a living being may be part and with which he may be in a relationship of mutual synthesis.'[60] The individual's health is, therefore, dependent on the health of the family, which is in turn dependent on a healthy social environment. A healthy person 'enriches his life and the social and physical environment he shares with others'.[61] Equally, the health of the individual is dependent upon a healthy diet, which requires food grown in a healthy soil, which requires a different attitude towards agriculture. Only a society which cares for its countryside and rural economy can be deemed a healthy 'body politic'; only a healthy body politic can produce healthy citizens.

We are thus led to the application of the organic metaphor to a society or state. This is one of the most problematic aspects of the organic viewpoint, since the metaphor's implications show a strong tendency towards totalitarianism. The German Romantic political philosopher Fichte claimed 'to be first to apply the simile of an organism to the whole civic relation... In the organic body every part continually maintains the whole, and while it maintains it, is itself maintained thereby; just such is the citizen's relation to the State.'[62] The ecologist Haeckel also wrote in such terms, describing the state as 'a unified organism of the higher order... for which [individuals] have no value except as parts of a whole'.[63] A British admirer of Italian fascism, H.E. Goad, wrote of 'the Corporate State [and] its embodiment as in a living triumphant organism';[64] and the organic principle of the priority of the whole can be seen in his assertion that 'According to Fascist theory the State is far from being merely "the sum of the citizens" that are its subjects...'.[65] E.I. Watkin cites Smuts as a major influence on fascist thought[66] and sees both fascism and communism as examples of the organic state. Historical links between fascism and the organic movement have been studied by Anna Bramwell and Richard Griffiths.[67]

On the other hand, it has been argued that it is the mechanistic viewpoint, not the organic, which involves the dangers of totalitarianism. 'Organic integration... begins from the realisation that the part reflects the whole and that the whole is mirrored in the parts; that the contribution of the part matters in relation to the whole; and that the integrity of the whole cannot be maintained at the expense of the integrity of the parts.'[68] For H. J. Massingham, the totalitarian states were the culmination of a value system that had rejected an organic outlook;[69] and the influential American ecologist Aldo Leopold was repelled by the Nazis' methods of land management.[70]

More recently, Wendell Berry has argued that the organic outlook is, by its very nature, opposed to totalitarian attitudes. Where agriculture is concerned, the industrial approach is 'a totalitarian form of behaviour... in its use of people as it is in its use of nature. Its connections to the world and to humans and the other creatures become more and more abstract, as its economy, its authority and its power become more and more centralized'.[71] Green politics can be traced back to de-centralist forms of socialism[72] as well as to fascism, and certainly there is a pluralist aspect to the organic movement which is just as important as the more sinister totalitarian side.

This is an enormously complex issue, touched on not because any solution is proposed, but because no outline of the concept of the organic can be complete without some reference to it. Can the recognition of variety and interconnectedness within an all-embracing hierarchy of wholes be reconciled with individual freedom, or will identity be subordinate to the requirements of a dictatorial co-ordinating authority? Duane Elgin has argued that the problem will be to balance *integration* and *differentiation* within an 'evolving social organism... The creative tensions between these two evolutionary vectors... require us to search continually for a skillful [*sic*] middle path that avoids the excess of either extreme (achieving integration at the expense of diversity or achieving diversity at the expense of integration).'[73]

For the philosophers Lossky, Watkin and Macmurray, the organic metaphor is far more adequate to the facts of experience than the mechanistic, but ultimately fails to do justice to human personality. When Lossky speaks of the 'super-organic' he means by this phrase the realm of spirit, or the Kingdom of God, which 'has the character of complete wholeness... [In it] each entity has... a definite function, so that through realizing its own unique nature completely and rightfully it manifests its super-individual significance...'[74]—that is, its significance for the whole. Watkin asserts: 'A society can be at once organic and free, only if it transcends the state';[75] and it can do this only if it recognizes something in its members which transcends material existence.[76] For Macmurray, the organic metaphor is inadequate because it cannot 'represent the nature of rational or objective consciousness as we know it in immediate experience... Human consciousness is not organic'.[77]

The organic points beyond the organic. An organic future—with its sense of the whole, its recognition of interconnections, of the aesthetic, and of variety—must be leavened by an awareness that humans are more than biological, and that to see their societies as analogous to the arrangements of other species is itself a form of reductionism, with dangerous implications.

Chapter Nineteen

Earth and Spirit—A Tradition Renewed

Robert Waller

MANY GREENS HAVE TURNED TO ORIENTAL RELIGIONS as more expressive of their intuitions than Christianity. The Christian religion has been severely criticized by conservationists—myself included—for treating nature as if it existed solely for the material benefit of mankind, even if mankind is promoted to being its steward. Stewards are tempted to regard their function as primarily to make a profit out of their assets like city councillors. Christians have found sanction for this in the creation myth of Genesis. This myth is, however, ambiguous and has two different sources. In any case, Christianity should be based on the teachings of Christ and these are often critical of Old Testament teaching and law, which on one occasion Jesus referred to as 'dead men's bones'. The Church has, however, as often been guided by dead men's bones as by the more vital and contemporary words of their founder. This has been especially true of some of the dissenters and non-conformists in the 19th century, who, contradicting the words of Jesus, taught that God blesses virtue with prosperity—so the prosperous are the good and virtuous—and that those who fail to prosper must have been sinners or layabouts.

This Old Testament Puritanism, with its so-called Puritan work ethic, has had for its motive the need for a creed that inspired and justified the accelerating economic growth of the industrial revolution. It has always been virtuous to be 'industrious'; it is, however, no longer the industry of the farm worker and craftsman that is rewarded with divine favour, but the monotonous work of the machine operator and the relentless planning of the financier and politician. The words 'work' and 'industrious' have been degraded. The association of this degradation with Christianity has turned many people, like myself influenced by the Green attitude of mind, against Christianity and made us seek elsewhere for a religion that satisfies our beliefs. At the same time, I have been aware that this puritanical Christianity has no relationship to the teachings of Jesus such as The Sermon on the Mount. Wouldn't it be better, therefore, to rediscover Christianity with fresh eyes as the Green religion that its founder intended that it should be?

The contrast between the teachings of Jesus—and the life of Jesus—and the way the industrialized nation states conduct their affairs has always inspired miracles of rhetorical sophistry to harmonize them with the teaching of their professed religion.

H.J. Massingham, in his book *The Tree of Life,* calls Chapter Two 'The Rural Christ'. Before this is rejected as absurd just consider the facts of the gospels. According to the Christian record of the life of Christ, God sent his Son to earth to redeem us. Where did he choose to send him? To be fathered by an emperor and empress in a palace? To be born in the villa of a latifundian agricultural magnate? Or the official residence of a provincial governor? Or perhaps to be the bastard of a high priest? No, he chose a peasant woman married to a village carpenter. The only city that the Son of God seems to have visited was Jerusalem, where he launched a fierce attack on the priestly establishment. His disciples were all rural people; his parables were mostly illustrated by episodes from rural life.

In other words it is hard to find any great teacher or prophet more green than Jesus—or any religion less green than Christianity, the religion of the industrial revolution that has subordinated the rural to the urban all over the world—with often appalling consequences. When asked what he thought of Christianity, Gandhi—who was greatly influenced by Jesus—asked: 'When are you going to try it?' A good question and one which the Greens can answer with: '*We* are going to try it'.

If Jesus is the Son of God, then God must be green as well and prefer peasant women and small craftsmen to emperors, tycoons, bankers and the like, unless they recognize the spiritual damage that they do through the kind of societies that they create. The treatment of money as a divine substance (yet one that is traded like any other commodity) is a deadly sickness. It is surely not surprising if God, the creator of nature, believes there is more wisdom, truth and honesty in rural life than in great cities; more understanding of life to be found in craftsmen and farmers—as farmers used to be, that is—than in more prestigious and wealthy occupations. And this is what the Greens are rediscovering today without realizing they are excavating Christianity and bringing it back to life. It explains why the rural is being recognized as the foundation of the stable civilization and yet is being destroyed.

When economists and politicians fail us, as they have done in impoverishing and disrupting the under-developed countries with their absurd creed of constant economic growth powered by unregulated, free, international trade, we must turn to poets and philosophers to restore wisdom. For they, participating in the life and spirit of nature, suffer with the destruction caused to the soul and the senses by the maladies of progress.

In 1929, Rabindranath Tagore, the Indian poet and philosopher who was concerned with the full realization of life, said: 'The wind from the West has scattered the seeds of social dissonance all over the world, destroying not only peace and happiness, but the very core of life itself. Here is a problem on which people everywhere must ponder.'

Indeed we must. Tagore pondered it for a lifetime and founded an Indian village community in the hope of reviving the culture and social life that had been destroyed. He was appalled by the dullness of village life deprived of its traditional culture; also, he saw that the British had increased the gap between the rich and the poor. He said of his village settlements: 'Our object is to try and flood the choked bed of village life with streams of happiness. For this the scholars, the poets, the musicians, the artists have to collaborate and offer their collaboration. Otherwise they live like parasites sucking life from country people and giving nothing back. Such exploitation gradually exhausts the soil of life.'

It has been observed by some American psychologists that the 'flow' of life has tended to dry up in Western communities. This is one of the maladies of progress; it has choked the natural streams of happiness. We have no real village life left; Tagore's observation applies just as much to our suburban life, in which spontaneous feeling and emotion are subordinated to economic fears and obsessions.

Tagore and Gandhi were both sophisticated, intellectual and well-travelled men—Gandhi was not at all like the ignorant fanatic that Churchill sneered at (thus exposing the coarse grain of materialism that soiled his courage and genius). Tagore and Gandhi maintained that, all over the world, the crisis at the root of both the urban and rural disorders resulted from the increasing greed of mankind and the ruthless exploitation of the strong by the weak.

This, in turn, is the consequence of the recognition by Nietzsche that 'God is dead'. Actually, Nietzsche is dead and God still lives, though hidden by the godless way we live. We have only to look around a modern city to see that. The temples are banks. If we are to recognize God as the creative spirit incarnated in the green Christ, then Western materialism has almost eliminated him. After all, Jesus *was* a villager. Massingham, in his book, draws out the disturbing implications of this for industrial, urban man.

Since Jesus chose the lowly status of a carpenter's son in this world, the Kingdom of God would hardly be managed by millionaires—'It is easier for a camel to go through the eye of a needle than a rich man to enter the Kingdom of Heaven'. It simplifies things so much when the things that are God's become the things that are Caesar's; it enables the industrial progress to drive the peasants off the face of the earth as it has already extinguished most of the hunter-gatherers; though in many parts of the world, where the peasants still exist in large numbers, they are putting up a bloody fight for their survival which receives little sympathy from the Christian West. They have forgotten Jesus was a peasant. The greatest peasant of all forbade his followers to use force; but we should never have put the world's peasantry into a situation in which they were tempted to use it in the hope of winning justice.

The Greens' personal philosophy (eco-organic) keeps their thinking close to nature, agriculture and the management of the earth's resources. It is, therefore, close to rural life, which it realizes is the foundation of religion, culture and civilization. With this outlook, it is in agreement with the founder of Christianity. Had the Greens had the power of the capitalist and communist Governments that have endeavoured to establish, in their different ways, the industrial society, we should still have the hunter-gatherers, prosperous rural societies and the unravaged rainforests, the fertile soils and the hedges and trees of well-farmed landscapes. We can only hope that the Green movement has not come too late to save our land and our souls in accordance with the teachings of the rural Jesus. The Greens should become the true Christians of today. They should awaken to this fact and, at the same time, dissociate themselves from the pseudo-Christianity of the politicians—as indeed they are doing. But the public does not yet realize the significance of this repudiation. There are signs, however, that some influential members of the Church are beginning to understand it— which is creating a crisis in ecclesiastical unity. This is inevitable if Christianity is to be reborn according to the teachings of Jesus. The teachings are more important than the unity of the Church.

As Massingham points out, Jesus was born in a manger among cattle, adored by shepherds, and received homage from wise men who did not scorn cattle and peasants; this is a story created with hindsight but embodying a profound truth. The wise men would, in time, realize that the conditions of Jesus' birth created a vision of a new society, one in which shepherds, fishermen, carpenters, farmers, animals and the earth's products would become symbols of the way humankind should live. Jesus' mind and imagination used rural occupations, rural traditions, rural proverbs and rural oral traditions, including poetic, tale-telling ways of talking, far removed from pedantic intellectualization, statistical analysis and economic theories elaborated in a disembodied void. He spoke always in terms of the personal life we all have to live and placed it within the pattern of a divine order. Consider, for example, the Parable of the Sower, in which three kinds of personality are compared to three kinds of soil. To ask oneself 'What kind of soil am I?' is a very organic way of self-analysis.

The rural society to which Jesus belonged had endured a profound religious history; it had survived hundreds of years of hope, inspiration and bitter experience—of which the people were well aware as they lived under the Roman yoke. Their Messiah was expected to free them from foreign domination; instead of that, Jesus looked inward and offered all mankind, including the Romans, a sustainable way of life. In the outcome, it was the Romans who adopted Christianity and his own establishment who rejected it. He did not support the sword to protect the Kingdom of Heaven; it is a very different sort of kingdom that does that: the kingdom of nuclear weapons; the kingdom of anti-Christ, as the heretics would have said in the Middle Ages.

There are complexities we have to disentangle and whose interpretations we have to leave open. Among the most difficult to interpret are the apocalyptic sayings of Jesus that the end of the world was at hand. In fact, this doom did not literally threaten his world as much as it does ours—with nuclear weapons, industrially-induced climatic changes, over-population, erosion of soils and the uncontrolled expansion of great insanitary cities. To think 'as if' the end of the world threatens sharpens our moral and intellectual insights until they come close to those of Jesus. Whether Jesus symbolized the end of the world—which happens to all of us in death anyway, so we can't take our wealth and power with us—or meant it literally it is hard to say and perhaps does not matter, so far as the basic truths of his teachings are concerned. The peasant has less to lose than the richer classes, so he is disinclined to fight imperial wars that leave the land abandoned and ravaged for the sake of loot that will never benefit him anyway and will probably be used to drive him off his land to make way for the vast agro-industrial estates which usually follow economic 'progress'. I shall never forget seeing German peasants ploughing with horses during the last war while we were fighting all around them, apparently unconcerned with such nonsense. It left a Hardy-like image of the eternal ploughman in my memory and seemed, somehow, also an image of Christian values.

The peasant and the rural craftsmen learn how to use with the optimum efficiency what nature provides for them; the peasant does not rob the world of its irreplaceable resources to enrich himself at the expense of others. Many of Jesus' parables urge us to enrich ourselves by recognizing we are spiritual beings and developing our spiritual talents. At the same time, being rural, he recognizes these talents develop through the crafts and activities of rural life—including fishing. He 'fishes' for souls. The spiritual tradition is a 'vine'; stone and water and wine, as in the Marriage at Cana, are potent symbols which every Christian ought to understand. The 'marriage' is the transformation of the stone and the water into the personal, spiritual wine of the spirit which changes the Self, so there is a new 'master of the house'. The marriage at Cana is a parable describing spiritual transformation from the dogmatic truths (the Ten Commandments were graven on stone) to the living truths of the gospels (water is the symbol of truth) to the final wine of love, the ultimate consummation. We have several levels of being and to integrate with the highest requires 'rebirth' through self-discipline and understanding. It does not happen automatically just because we are alive. But if the truly Christian life is a tradition and an ideal taught from childhood, it makes wisdom easier to achieve.

The Greens should aim at this transformation of being taught by Jesus as essential to creating a sustainable society by a double revolution—an inward revolution of the Self coinciding with social reforms. The Greens must have personalities capable of resisting the temptations of the world. 'Be of good cheer; I have overcome the world', as Jesus said. Liberation from the world releases a great joy and happiness such as Jesus clearly meant by the Kingdom of Heaven. It makes poets. It also makes objective thinkers who do not judge everything according to how it advances their personal careers and gratifies their egotism.

This green teaching of Jesus was understood by Shakespeare. Consider what Albany says of Lear's evil daughter Goneril in *King Lear*:

That nature, which contemns its origin,
Cannot be bordered certain in itself;
She that herself will sliver and disbranch
From her material sap, perforce must wither
And come to deadly use.

The fatal error of industrial civilization is the extent to which it has disbranched itself from its material sap, so the mind has become disembodied and abstract, able to imagine the most appalling projects—such as cutting down all the trees—as productive enterprises; projects which, did we live natural lives, we would recognize at once as evil, as the peasants in Nepal and in Africa do who rope themselves to their trees.

'And why do you worry about clothes? See how the lilies of the field grow. They do not labour or spin. Yet I tell you that not even Solomon in all his splendour was dressed like one of these. If that is how God clothes the grass of the field, which is here today and tomorrow is thrown into the fire, will he not much more clothe you, O you of little faith? So do not worry, saying "What shall we drink?" or "What shall we wear?"... But seek first God's kingdom and his righteousness, and all these things will be given to you as well. Therefore do not worry about tomorrow, for tomorrow will worry about itself. Each day has enough troubles of its own.'

It seems to me that has a green flavour about it. These are not the words of a man who spends his time worrying about his investments, consulting the financial columns of the papers every day, investing in the destruction of forests and the commercialization of water, and trying to anticipate the next merger so he can make a few thousand on the shares. Where a man's treasure is there will his heart be also. You cannot love God and Mammon.

Let us consider it ourselves, as a guide to practical, economic farming. The outstanding threat to the survival of our vast cities and huge populations is soil erosion. The protector of the soil, above all others, is undisturbed vegetation, particularly the prairie or what we call permanent grass. Whenever the soil is disturbed by ploughing, some invaluable soil is lost. This exposure of the soil to loss in the process of husbandry by turning the land wrong side up is increased, one might almost say—world-wide—millionfold, by modern farming, because it does not return to the soil what it has taken from it in the form of natural clothing, that is, mulch. If, therefore, the cultivator were to pay more attention to, and have more respect for, the way nature (or God) provides a sustainable fertility, if he refrained out of greedy anxiety about tomorrow from ploughing up and destroying the natural structure of soils, he would then realize that the problem of how to feed the multitudes and keep nature's productivity at an optimum indefinitely, would be addressed by directing research into securing the secrets of nature, rather than teaching nature how to perform its own operations. The Sermon on the Mount conveys this message to Christians in the poetic way in which Christ always spoke; he would only have spoken ironically of nature's 'productivity machine'—the sort of language that we use and which strips nature of enchantment, sacredness and truth, and, in the end, of productivity as well.

Jesus, by identifying himself with nature and God and their creativeness, and attaining this vision by transcending his own egoistic individuality, created a green religion of living that, now I come to reflect on it in my old age, seems to me to offer as much ecological wisdom as other religions.

From fear of being misunderstood I must make one last observation. For most of its history, human society has been predominantly rural; in this rural world prophets and teachers created great religions and sought for wisdom; gradually cities, too, have made their contribution to the development of the human faculties and the creation of wealth. It is a truism that the city, more than any other environment, stimulates the critical faculties. But, since the industrial revolution, the creation of urban societies has seemed the only way to salvation and a sustainable economy capable of meeting the needs of expanding populations. This belief has taken such a hold that the unindustrialized societies, in order to compete, have driven the peasants from the land, razed villages and forced the rural population to live in towns. This is an error. The only way to create a sustainable society and economy is to create a proportionate balance between urban and rural. In a balanced world, in which the rural have access to the urban and the urban access to the rural, so that the best qualities of both can converge, the truths of religion revealed in the rural past can still mould the way people live, but they cannot in a wholly urban world. The close cultural interpenetration of the rural and the urban will lead to opportunities for a more complete life. We will have to ruralize the urban as well as urbanize the rural. Indeed, in this last respect, the industrialized world is now over-developed. Its great days of excess affluence are over. It is no longer morally safe. The future of the organic movement lies in the realization of this. The many spontaneously-arisen non-governmental-organizations (NGOs) are compelling governments world-wide to take account of the personal needs of citizens who are refusing to be enslaved by overbearing economic policies that take no account of the more intimate desires and feelings of the individual, whether these are coercive socialist policies or anarchic, market-dominated, capitalist policies. The traditional teachings of Christianity put the salvation of the soul first. How is the soul saved? Not simply by muttering a dogmatic belief, but by faith in the creative powers of the person to live a complete life in an intuitively felt divine pattern of creation.

NOTES ON CONTRIBUTORS

Michael Allaby was on the staff of The Soil Association from 1964 until 1972, becoming assistant editor of its quarterly journal and editor of its monthly magazine. During this time he was actively involved in the Association's rescue of the Conservation Corps and its re-launching as the British Trust for Conservation Volunteers, and in *The Ecologist* magazine, which for a time was produced by the Soil Association editorial department. In 1972, *The Ecologist* moved to Cornwall and he moved with it as its managing editor. Since 1973 he has been a freelance author. He is editor of the *Macmillan Dictionary of the Environment*, *The Oxford Dictionary of Natural History*, *The Concise Oxford Dictionary of Zoology*, *The Concise Oxford Dictionary of Botany* (in press) and, with Ailsa Allaby, of *The Concise Oxford Dictionary of Earth Sciences.* He is editor of the new series of official guides to all the national trails in England and Wales and has also written or co-authored more than 30 books on natural history, popular science, and the environment.

Graham Bell is editor of *Permaculture News*, quarterly magazine of the British Isles Permaculture Institute. He is a designer of land use, small-scale enterprise and horticultural systems. In 1990 he was awarded the Permaculture Community Service Award by the International Permaculture Institute. He lives in a small community in the Scottish Borders. He has published articles on a wide range of subjects and a book on *Temperate Climate Land Use Design*, published by Thorson's in Autumn 1991. He has previously worked in energy efficiency, computing and communications, and in the construction industry. He is a gardener by desire.

Will Best was educated privately and at Trinity College, Cambridge. He took on the tenancy of his father's Dorset farm after his marriage in 1969, at the age of 21, and concentrated on modernizing it and making it viable. Conversion to organic philosophy and practice came later. He and his wife Pam have three children, and a variety of commitments in the local community. Pam's grandfather founded and ran the Dorset branch of CPRE, and Will's father revived it with the estate-owner Rolf Gardiner in the 1950s.

Sir Richard Body practised as a barrister for 20 years, from 1949, and has been Member of Parliament for Holland-with-Boston since 1966. His concern for agricultural policy stems from his own experience as a farmer whose stock has won numerous championships and been exported to China, South America and the East Indies. He has been a judge at the Royal Show and has undertaken round-the-world lecture tours on agriculture. From 1974 to 1987 he was a member of the Parliamentary Select Committee on Agriculture, being elected Chairman in 1986. He has written many articles on farming policy, and four books, the most recent of which is *Our Food, Our Land* (1991).

Peter Bunyard read Natural Sciences at Cambridge and studied insect physiology at Harvard, gaining MAs from both universities. From 1966 to 1972 he was Science Editor of *World Medicine* and, in 1969, he became founding co-editor of *The Ecologist.* Since 1977 he has been Consultant Editor of the UNEP's *Industry and Environment Review.* He is the author of many articles in *The Ecologist* on energy and on the tropical forests, and of several books, including *The Green Alternative Guide to Good Living* (co-author/editor, 1987); *A Health Guide for the Nuclear Age* (1988); *The Colombian Amazon* (1990); and *5,000 Days to Save the Planet* (co-author, 1990). He is a member of the faculty of the Global Ecology Course—the International Honours Programme; and has given seminars at the Schumacher College on the Gaia Hypothesis and on Ancient Wisdom in Modern Times.

Philip Conford graduated in philosophy at the University of East Anglia, where he also took an MA in comparative literature. He has taught English since 1974, first in the penal system and subsequently in further education. He is the editor of *The Organic Tradition* (Green Books, 1988), and has contributed to *Resurgence* and the *Times Higher Education Supplement.*

Ron Davies was elected to the House of Commons in 1983, after spending 14 years in local politics as a district councillor, the last 5 as deputy leader of Rhymney Valley District Council and Chair of the Finance and Policy Committee. His parliamentary seat, Caerphilly, in the Rhymney valley, is largely rural, containing predominantly mixed and dairy farms.

He was appointed to the Opposition Whips' Office in 1984, where his subject areas included agriculture and the environment. Since 1987 he has been in Labour's front-bench agriculture team as deputy to the Party's senior food and agriculture spokesman.

His life-long interest in the countryside, wildlife and the environment has been put to good use in helping reconstruct Labour's policies in these areas since 1987. He considers his interest in organic agriculture to be a natural corollary to his commitment to environmentalism, as it applies to land use and food production.

Penny Evans is a graduate of the Forestry department of the University of Edinburgh and has worked principally in the field of nature conservation. She has been responsible for the purchase and management of a number of woodland nature reserves but now works in the policy field. She is currently employed by the Council for the Protection of Rural England, where she is responsible for transport policy as well as forestry policy.

Koyu Furusawa was born in 1950 and studied biology at Osaka University, also participating in Open Forum anti-pollution activities. Deeper involvement in environmental groups led to a concern for agriculture, and he undertook graduate studies in agricultural economics, as well as working for the Society for Reflecting on the Throwaway Age. Since 1982 he has been involved with the Seikatsu Club co-operative organization and has travelled widely in South-east Asia and the USA, establishing contacts with activities and researchers concerned with global problems in agriculture, environment and economics. He organized the International Symposium on Rice and Food Self-Sufficiency held in Tokyo in 1988. He is co-author of *Workers' Collectives*, and is himself a member of a collective specializing in educational preparation. He teaches at various universities and writes on the relation between agricultural problems and world-wide economic forces.

Alan Gear is Chief Executive of The Henry Doubleday Research

Association at The National Centre for Organic Gardening. He has lectured and broadcast widely on the subject of organic gardening, and is best known as presenter of the highly successful and widely acclaimed Channel 4 TV series *All Muck and Magic?*

He has written many newspaper and magazine articles, as well as information booklets and contributions to the HDRA newsletter. In 1983, he wrote *The Organic Food Guide* for HDRA. This was updated and reprinted as *The New Organic Food Guide* (Dent, 1986). He co-edited *Thorsons Organic Consumers Guide* (Thorsons, 1990), and was a contributor to *Thorsons Organic Wine Guide* (1991). He was a founder member of the British Organic Standards Committee, and is now regarded as a central figure in the organic movement.

Professor Richard Grantham was born in Kingsburg, California, in 1922. During World War II he was a USAAF Liberator pilot in Europe. In 1953 he took a BA degree in Chemistry at the University of Southern California. From 1948 to 1965 he was a research scientist in the petroleum and aerospace industries. In 1966 he began research on molecular evolution in France, completing his *Doctorat d'Etat ès Sciences* in 1975, and subsequently becoming Professor of Biology at the Université Claude Bernard, Lyon. In 1982 he was appointed Director of the Institut d'Evolution Moléculaire. His personal research since 1986 has been devoted to future evolution and global ecology, especially with reference to possibilities for correcting the greenhouse drift. Since 1989 he has been Chairman of the working group on 'Geotherapeutic Approaches' of the Committee on Global Change of INQUA (International Union for Quaternary Research). Since October 1st 1990 he has been Professor Emeritus at the Université Claude Bernard. He is married and has three children.

Wes Jackson, President of The Land Institute, was born in 1936 on a farm near Topeka, Kansas. After attending Kansas Wesleyan (BA biology, 1958), he studied botany (MA University of Kansas, 1960) and genetics (PhD North Carolina State University, 1967). After a short stint of teaching at high school, he taught in the biology department at Kansas Wesleyan and later was hired to establish the Environmental Studies programme at California State University, Sacramento, where he became a tenured full professor. He resigned that position to found The Land Institute in 1976 near Salina, Kansas. He has written numerous papers and book chapters and has four book titles. One which he edited, *Man and the Environment*, 1970, went through three editions and was widely adopted as a text. *New Roots for Agriculture,* published in 1980, is the basis for the agricultural work at The Land Institute. *Meeting the Expectations of the Land*, edited with Wendell Berry and Bruce Colman, appeared in 1984. His last book, *Altars of Unhewn Stone,* was published in 1987. The work of The Land Institute has been featured in *The Atlantic, Audubon*, and National Public Radio's 'All Things Considered', among others. *Life* magazine named Wes Jackson as one of 18 individuals they predict will be among 100 of the 'important Americans of the 20th century'.

Wangari Maathai is the founder and coordinator of the Green Belt Movement. She was born in Nyeri, Kenya and has a Bachelor of Science degree from Mount St Scholastica College, Kansas, USA; a Master of Science degree from the University of Pittsburgh; and a PhD from the University of Nairobi, Kenya, where she was Professor of Anatomy. She is the mother of three children.

Robert I. Papendick is head of the Land Management and Water Conservation Research Unit with the United States Department of Agriculture (USDA) Agricultural Research Service in Pullman, Washington. He served as chairman and coordinator of the study *Report and Recommendations on Organic Farming*, published in 1980.

James F. Parr is a soil-fertility programme leader with the USDA Agricultural Research Service in Beltsville, Maryland, and an authority on crop-residue management systems for soil and water conservation.

John P. Reganold teaches introductory soil science and conservation and management at Washington State University, where he is Associate Professor. He has conducted several studies that compare the effects of conventional and organic farming methods on soil systems.

Fiona Reynolds graduated from Cambridge University in 1979 with an Honours degree in Geography and Land Economy. The following year, she was awarded an MPhil by the Department of Land Economy at Cambridge University and was appointed secretary to the Council for National Parks, a post she held till 1987. Since then, she has been Assistant Director (Policy) for the Council for the Protection of Rural England. In 1990 she was given the 'Global 500' Award from the United Nations Environment Programme (UNEP), for outstanding achievement in protection and improvement of the environment. She is a member of the National Parks Review Panel.

Diana Schumacher is a partner in a business management consultancy. For the past 17 years she has also been working on the environmental implications of energy use and abuse. She is an active participant and committee member of numerous environmental organizations, including the New Economics Foundation, the Ecological Action Group for Europe (ECOROPA), Church of England Environmental Reference Panel, The Churches' Energy Group, The Parliamentary Alternative Energy Group (PAEG), The Environmental Law Foundation (ELF) and the Schumacher Society. She has published many reviews, books and articles on alternative energy and the environment, with a particular emphasis on the developing countries and grass-roots needs.

Vandana Shiva is director of the Research Foundation for Science, Technology and Natural Resource Policy, Dehra Dun. She is a physicist, a philosopher of science, and an ecologist. She has been active in citizens' movements against environmental destruction, including the Chipko movement, and at present is particularly concerned with the preservation of diversity, which she sees as a condition for continued existence. She is the author of the books *Staying Alive*, *Violence of the Green Revolution*, *Ecological Audit of Eucalyptus Cultivation*, and *Ecology as the Politics of Survival.*

Robert Waller edited the Soil Association Journal from 1963 to 1972. He is the author of *Be Human Or Die*, a philosophical study of human ecology, and the biographer of Sir George Stapledon, the agrarian scientist. He is joint editor of the Commonwealth Human Ecology Journal, and also a poet and novelist. He has been a regular contributor to the Soil Association's journal, *Living Earth.*

REFERENCES

Introduction

1. Cf. 'Organic Farming and the Environment' by Joy Greenall, in *Organic Farming—An Option for the Nineties* (Bristol, British Organic Farmers/ Organic Growers Association, 1990).

2. Barry Wookey, *Rushall—The Story of an Organic Farm* (Oxford, Basil Blackwell, 1987) pp.42-43.

3. Ibid. p.44.

4. Cf. Chapter 6 of Philip Conford (ed.) *The Organic Tradition* (Hartland, Green Books, 1988); *The Peckham Experiment*, By I.H. Pearse and L.H. Crocker (London, Allen and Unwin, 1943); *Thoughts on Feeding*, by L.J. Picton (London, Faber, 1946); and *Being Me and Also Us*, by Alison Stallibrass (Edinburgh, Scottish Academic Press, 1989).

5. Cf. Rolf Gardiner, *England Herself* (London, Faber, 1943); H.J. Massingham (ed.), *The Natural Order* (London, Dent, 1945); The Earl of Portsmouth, *Alternative to Death* (London, Faber, 1943).

6. Cf. H.J. Massingham, *The Faith of a Fieldsman* (London, Museum Press, 1951)—particularly 'The Bargain Counter', 'Terrace Cultivation', 'African Survey', 'The Dark Continent', and 'The Felling of the Broad-Leaved Forests'.

7. Cf. H.J. Massingham, *Remembrance* (London, Batsford, 1942), as representative of the organic husbandry movement, or, among contemporaries, John Seymour's *The Ultimate Heresy* (Hartland, Green Books, 1989).

8. *Daily Telegraph*, Gardener Supplement, (4 July 1989), p.5.

9. Marion Shoard, *The Theft of the Countryside* (London, Maurice Temple Smith, 1980). Chapters 3–9 deal with landscape features.

10. Masanobu Fukuoka, *The Road Back to Nature* (Tokyo and New York, Japan Publications Inc., 1987), p.330.

11. *Living Earth*, January–March 1990, p.4.

12. For a discussion of the debate, see Tim Cooper, *Green Christianity* (London, Spire, 1990), pp.33–38; and Jonathon Porritt and David Winner, *The Coming of the Greens* (London, Fontana, 1988), pp.241-245.

13. Edward Goldsmith, Nicholas Hildyard, Peter Bunyard and Patrick McCully, *5,000 Days to Save the Planet*, (London, Hamlyn, 1990).

Chapter 6: A Policy for Forestry and Woodland

1. *Land Use and Forestry*, Vol. 1. House of Commons Agriculture Committee, 1989.

2. *Budgeting for British Forestry* (London, Pieda/CPRE, 1987).

3. Steve Tompkins, *Forestry in Crisis—the Battle for the Hills* (Bromley, Christopher Helm, 1989).

Chapter 11: Towards the Marriage of Ecology and Economics

1. J. Vladislav, ed., *Vaclav Havel: Living in Truth* (London, Faber, 1989), p.136.

2. Ibid. p.138.

3. Ibid. p.139.

4. Ibid. pp.139–40.

Chapter 12: Indigenous Rights—Colombia's Policy for the Amazon

1. German Andrade and Juan Pablo Ruiz, *Amazonia Colombia: Aproximación ecológica y social del la colonization del bosque tropical*, FESCOL 4, 1988.

2. *Politica del Gobierno Nacional para la Defense de los Derechos indigenas y la Conservacion Ecologica de la Cuenca Amazonica*, Republica de Colombia, November 1989.

3. Tomas Walschburger and Patricio von Hildebrand, 'Observaciones sobre la Utilizacion Estacional del Bosque Humedo Tropical por Los Indigenas del Rio Miriti', *Colombia Amazonica*, Vol.3, No.1, 1988.

4. Tomas Walschburger and Patricio von Hildebrand, art. cit.; Ana Walschburger 'Algunos aspectos generales sobre las repercusiones ecológicas del systema de tumba y quema de los indigenas Yacuna en la Amazonia colombiana', *Colombia Amazonica*, Vo.2, No.2, 1987.

5. Christopher Uhl and Juan Saldarriaga, *Scientific American* No.121 (October 1986).

6. Tomas Walschburger and Patricio von Hildebrand, art.cit.

Chapter 13: Recovering Diversity—A Future for India

1. C. Caufield, *In the Rainforest* (London, Picador, 1986), p.60.

2. E. Hong, *Natives of Sarawak* (Malaysia, Institut Masyarakat, 1987), p.137.

3. S. C. Chin, *The Sustainability of Shifting Cultivation* (Penang, World Rainforest Movements, 1989).

4. J. de Beer and M. McDermott, *The Economic Value of Non-Timber Forest Products in Southeast Asia* (Amsterdam, Netherlands Committee for IUCN, 1989).

5. V. Shiva, *Staying Alive* (London, Zed Books, 1988), p.59.

6. Both Grigson and Tiwari are quoted in M.S. Randhawa, *A History of Agriculture in India* (New Delhi, Indian Council of Agricultural Research, 1980), p.99.

7. Ibid. p.97.

8. K.K. Panday, *Fodder Trees and Tree Fodder in Nepal* (Berne, Swiss Development Cooperation, 1982).

9. S.P. Singh and A. Berry, *Forestry Land Evaluation at District Level* (Bangkok, FAO, 1985).

10. T.B.S. Mahat, *Forestry—Farming Linkages, in the Mountains* (Kathmandu, ICIMOD, 1987).

11. S. Schlick, *Systems of Sylviculture* (1920).

12. R.S. Troup, *Sylvicultural Systems* (Oxford, OUP, 1916).

13. J. Bethel, 'Sometimes the Word is "Weed"' in *Forest Management* (June, 1984), 17-22.

Chapter 14: Co-operative Alternatives in Japan

1. Koyu Furusawa, 'Life Rooted in the Rice Plant', *Resurgence,* No. 137 (Nov/Dec 1989).

2. Paul Ekins, 'Growing Concern', *The Guardian* (13 January 1988).

3. Masanobu Fukuoka, *The One-Straw Revolution* (Emmaus, Rodale Press, 1978); *The Natural Way of Farming* (Tokyo and New York, Japan Publications, Inc., 1985); *The Road Back to Nature* (Tokyo and New York, Japan Publications, Inc., 1987).

Chapter 16: Towards an Organic Energy Policy

1. *Our Common Future* (UN World Commission on Environment and Development, OUP, 1987).

2. *Energy Policies and Programmes of IEA Countries, 1988 Review* (Paris, OECD/IEA, 1989), p.47.

3. 'The Green Ultimatum of Building Design', Transactions of a Conference held in Bristol in March 1980, Centre for Organisations Relating to the Environment, p.8.

4. J. Davis and A. Bollard, *As Though People Mattered: a Prospect for Britain* (London, Intermediate Technology Publications, 1986), pp.179-80.

5. World Action for Recycling Materials & Energy from Rubbish (WARMER), 83 Mount Ephraim, Tunbridge Wells, Kent TN4 8BS.

6. *The Cornwall Energy Action Plan*, Cornwall Energy Project, Report No. 7, February 1989.

7. Diana Schumacher, 'Energy for Human Shelter within the Global Shelter: Energy Policy, Planning for the Third World' in *Homes above All* (UK, Building and Social Housing Foundation, Russell Press Ltd., 1987).

8. Y. El Mahgary and A.K. Biswas, *Integrated Rural Energy Planning* (UK, Butterworths, 1985).

9. Carl J. Weinberg and Robert H. Williams, 'Energy from the Sun', *Scientific American* (September 1990).

10. C. Perrow, *Normal Accidents:—Living with High-Risk Technologies* (USA, Basic Books, 1984).

11. B. John, *The Global Impacts of Energy Technologies*, Internal Report No. 1, Energy Parks UK Ltd, Trefelin, Cilgwyn, Newport, Pembs SA42 0QN.

12. O. Hohmeyer, *Social Costs of Energy Consumption: External Effects of Electricity Generation in the Federal Republic of Germany* (Berlin, Springer-Verlag).

13. L. Brown *et al.*, *State of the World 1990*, A Worldwatch Institute Report on Progress Toward a Sustainable Society (Norton, New York, 1990).

14. Edward Goldsmith, Nicholas Hildyard, Peter Bunyard and Patrick McCully, *5,000 Days to Save the Planet* (London, Hamlyn, 1990).

15. H. Girardet, *Gaia* Magazine, Issue 2, 1990, p.10.

Chapter 17: The Healing of the Earth

1. R. Grantham, 'Approaches to correcting the global greenhouse drift by managing tropical ecosystems', *Tropical Ecology* 30 (2) (1989), 157-174.

2. M.J. Prather and R.T. Watson, 'Stratospheric ozone depletion and future levels of atmospheric chlorine and bromine', *Nature*, 344 (1990), 729-734.

3. S. Arrhenius, 'On the influence of carbonic acid in the air upon the temperature of the ground', *Philosophical Magazine* 41 (1896), 237-275.

4. J. Grinevald, 'L'effet de serre de la Biosphère de la révolution thermo-industrielle à l'écologie globale', *Stratégies énergétiques, Biosphère et Société*, 1 (1990), 9-34.

5. R. Grantham, 'Managing global change by curtailing emission sources and creating new sinks' in H.W. Scharpenseel, M.C.B. Shomaker and A.T. Ayoub (eds.) *Soils on a Warmer Earth* (Amsterdam, Elsevier, 1990) pp.221-230.

6. R. Grantham, 'Scientific and philosophical questions on development of a compensatory process for CO_2 buildup by greening of the Sahara' in A.M. Liquori (ed.) *The Ethics of Scientific Knowledge* (Rome, Enciclopedia Italiana, 1989), pp.103-116.

7. R. Grove, 'Origins of environmentalism', *Nature*, 345 (1990), 11-14.

8. J. Lelieveld and P.J. Crutzen, 'Influences of cloud photochemical processes on tropospheric ozone', *Nature*, 343 (1990) 227-233.

9. R.E. Newell, H.G. Reichle Jr. and W. Seiler, 'Carbon Monoxide and the Burning Earth', *Scientific American*, 262 (1) (1989), 82-88.

10. R. Grantham, 'Managing global change by curtailing emission sources and creating new sinks' in H.W. Scharpenseel, M.C.B. Shomaker and A.T. Ayoub (eds.) *Soils on a Warmer Earth* (Amsterdam, Elsevier, 1990) pp.221-230.

11. T.E. Graedel and P.J. Crutzen, 'The Changing Atmosphere', *Scientific American*, 261 (3) (1989), 58-68.

12. P.R. Crosson and N.J. Rosenberg 'Strategies for Agriculture', *Scientific American*, 261 (3) (1989), 128-135.

13. T.J. Goreau and W.Z. de Mello, 'Tropical deforestation: some effects on atmospheric chemistry', *Ambio*, 17 (4) (1988), 275-281.

14. N. Yoshida, H. Morimoto, M. Hirano, I. Koike, S. Matsuo, E. Wada, T. Saino and A. Hattori, 'Nitrification rates and ^{15}N abundances of N_2O and NO_3 in the western North Pacific', *Nature*, 342 (1989), 895-897.

15. D.A. Lashof and D.R. Ahuja, 'Relative contributions of greenhouse gas emissions to global warming', *Nature*, 344 (1990), 529-531

16. R. Grantham, 'Approaches to correcting the global greenhouse drift by managing tropical ecosystems', *Tropical Ecology* 30 (2) (1989), 157-174.

17. R. Grantham, 'Managing global change by curtailing emission sources and creating new sinks' in H.W. Scharpenseel, M.C.B. Shomaker and A.T. Ayoub (eds.) *Soils on a Warmer Earth* (Amsterdam, Elsevier, 1990) pp.221-230.

18. K.B. Hogan, J.S. Hoffman and A.M. Thompson, 'Methane on the Greenhouse Agenda', *Nature* 354 (1991), 181-182.

19. G.M. King, 'Regulation by light of methane emissions from a wetland', *Nature* 345 (1990), 513-515.

20. J. Chappellaz, J.M. Barnola, D. Raynaud, Y.S. Korotkevich and C. Lorius, 'Ice-core record of atmospheric methane over the past 160,000 years', *Nature* 345 (1990), 127-131.

21. D.A. Lashof and D.R. Ahuja, 'Relative contributions of greenhouse gas emissions to global warming', *Nature*, 344 (1990), 529-531

22. R.E. Newell, H.G. Reichle Jr. and W. Seiler, 'Carbon Monoxide and the Burning Earth', *Scientific American*, 262 (1) (1989), 82-88.

23. T.J. Goreau and W.Z. de Mello, 'Tropical deforestation: some effects on atmospheric chemistry', *Ambio*, 17 (4) (1988), 275-281.

24. S.C. Whalen and W.S. Reeburgh, 'Consumption of atmospheric methane by tundra soils', *Nature* 346 (1990), 160-162.

25. M.J. Molina, T.-L. Tso, L.T. Molina and F.C.-Y. Wang, 'Antarctic Stratospheric Chemistry of Chlorine Nitrate, Hydrogen Chloride and Ice:

Release of Active Chlorine', *Science* 238 (1987), 1253-1257.

26. R. Grantham, 'Approaches to correcting the global greenhouse drift by managing tropical ecosystems', *Tropical Ecology* 30 (2) (1989), 157-174.

27. I. Whitting, 'Natural gas', *New Scientist*, 12 May 1990, 77.

28. R. Grantham, 'Managing global change by curtailing emission sources and creating new sinks' in H.W. Scharpenseel, M.C.B. Shomaker and A.T. Ayoub (eds.) *Soils on a Warmer Earth* (Amsterdam, Elsevier, 1990) pp.221-230.

29. A.M. Hough and R.G. Derwent, 'Changes in the global concentration of tropospheric ozone due to human activities', *Nature* 344 (1990), 645-648.

30. J. Lelieveld and P.J. Crutzen, 'Influences of cloud photochemical processes on tropospheric ozone', *Nature*, 343 (1990) 227-233.

31. S.E. Schwartz, 'Chemistry with a silver lining', *Nature* 343 (1990), 209-210.

32. R. Stone, 'NRC Faults Science Behind Ozone Regs.', *Science* 255 (1992), 26.

33. D.A. Fisher, C.H. Hales, D.L. Filkin, M.K.W. Ko, N.D. Sze, P.S. Connell, D.J. Wuebbles, I.S.A. Isaksen. and F. Stordal, 'Model calculations of the relative effects of CFCs and their replacements on stratospheric ozone', *Nature* 344 (1990), 508-512.

34. D.A. Fisher, C.H. Hales, W.-C. Wang, M.K.W. Ko, and N.D. Sze, 'Model calculations of the relative effects of CFCs and their replacements on global warming', *Nature* 344 (1990), 513-516.

35. D.A. Lashof and D.R. Ahuja, 'Relative contributions of greenhouse gas emissions to global warming', *Nature*, 344 (1990), 529-531

36. P. Aldhous, 'Ozone report puts US policy in question', *Nature* 353 (1991), 783.

37. J. Hecht, 'Polluted atmosphere offsets ozone hole', *New Scientist*, 18 January 1992, p18.

38. M.J. Prather and R.T. Watson, 'Stratospheric ozone depletion and future levels of atmospheric chlorine and bromine', *Nature*, 344 (1990), 729-734.

39. R.M. White, ' The great climate debate', in *Scientific American* 263 (1) (1990), 36-43.

40. C. Kuo, C. Lindberg and D.J. Thomson, 'Coherence established between atmospheric carbon dioxide and global temperature', *Nature* 343 (1990), 709-714.

41. J. Chappellaz, J.M. Barnola, D. Raynaud, Y.S. Korotkevich and C. Lorius, 'Ice-core record of atmospheric methane over the past 160,000 years', *Nature* 345 (1990), 127-131.

42. T.J. Goreau, 'Balancing atmospheric carbon dioxide', *Ambio* 19 (5) (1990), 230-236.

43. R. Grantham, 'Castles in the Saharan air', *Nature* 325 (1987), 384.

44. R. Grantham, 'Scientific and philosophical questions on development of a compensatory process for CO_2 buildup by greening of the Sahara' in A.M. Liquori (ed.) *The Ethics of Scientific Knowledge* (Rome, Enciclopedia Italiana, 1989), pp.103-116.

45. R. Grantham, 'Cogene', *International Geosphere-Biosphere Report,* 7:2 (1989), 259-264.

46. R. Grantham, 'Approaches to correcting the global greenhouse drift by managing tropical ecosystems', *Tropical Ecology* 30 (2) (1989), 157-174.

47. A. Raval, and V. Ramanathan, 'Observational determination of the greenhouse effect', *Nature* 342 (1989), 758-761.

48. H. Flohn and A. Kapala, 'Changes of tropical sea-air interaction processes over a 30-year period', *Nature* 338 (1989), 244–246.

49. U. Mikolajewicz, B.D. Santer and E. Maier-Reimer, 'Ocean response to greenhouse warming', *Nature* 345 (1990), 589-593.

50. S. Manabe and R.T. Wetherald, 'Reduction in summer soil wetness induced by an increase in atmospheric carbon dioxide', *Science* 232 (1986), 626-628.

51. F. Berkes, D. Feeny, B.J. McCay and J.M. Acheson, 'The benefits of the commons', *Nature* 340 (1989), 91-93.

52. E. Brown Weiss, 'In fairness to future generations', *Environment* 32 (2) (1990), 7-31.

53. V.R. Potter, *Global Bioethics* (Michigan State University Press, 1988).

54. R. Grantham, Conference Report, 'Colloquium on Modelling and Geotherapy for Global Changes', *Global Environmental Change*, Vol 2 No 1 (1992) (in press).

55. R. Grantham, 'Evolutionary Choices in Geotherapy', *Global Environmental Change*, Vol 2 No 1 (1992) (in press).

Chapter 18: The Organic in Theory and Practice

1. Jeremy Naydler, 'Reviving the Forests', *Resurgence* No. 136 (September/ October 1989), p.52.

2. Lancelot Law Whyte (ed.), *Aspects of Form* [1951] (London, Lund Humphries, 1968), p.2.

3. L.J. Henderson, *The Order of Nature* (Cambridge, Mass., Harvard U.P., 1925), p.21.

4. Donald Worster, *Nature's Economy* [1977], (Cambridge U.P., 1985) gives a detailed account of different attitudes to nature since the 18th century, and is invaluable reading for an understanding of the concept of the organic. Similarly worthwhile are *The Turning Point* by Fritjof Capra [1982] (London, Flamingo, 1983); John Macmurray's philosophical study *Interpreting the Universe* (London, Faber, 1933); and *Organic Form—The Life of an Idea*, edited by G.S. Rousseau (London, RKP, 1972). My debt to these four books is considerable.

5. Vandana Shiva, *Staying Alive* (London, Zed Books, 1988) pp.40–41.

6. Donald Worster, op. cit. p.85.

7. W.K. Wimsatt, 'Organic Form—some questions about a metaphor', in G.S. Rousseau (ed.) *Organic Form* (London, RKP, 1972) pp.67–68.

8. Donald Worster, op. cit., p.82. For a more detailed account of Goethe's interest in organic forms, see the essay by Philip C. Ritterbush in *Organic Form*, ed. G.S. Rousseau (London, RKP, 1972).

9. Basil Willey, *Nineteenth-Century Studies* [1949] (Harmondsworth, Peregrine Books, 1964), p.38.

10. J.C. Smuts, *Holism and Evolution* [1926] (London, Macmillan, 1927) p.101.

11. Ibid. p.103.

12. N.O. Lossky, *The World as an Organic Whole* (London, Humphrey Milford/O.U.P., 1928), p.18.

13. L.L. Whyte, editorial preface to *Aspects of Form* [1951] (London, Lund Humphries, 1968), p.xi.

14. Rupert Sheldrake, 'Rebirth of Nature', *Resurgence* No.136 (September/October 1989), p.34.

15. Ronald Harvey, *Our Fragmented World* (Hartland, Green Books, 1988), p.2.

16. Wendell Berry, *The Unsettling of America* [1977] (San Francisco, Sierra Club Books, 1986), p.110.

17. Wes Jackson, *Altars of Unhewn Stone* (San Francisco, North Point Press, 1987), p.75.

18. Vandana Shiva, op. cit. pp.14-15.

19. Donald Worster, op. cit. p.58.

20. Ibid., p.82.

21. Ibid., p.82.

22. Ibid., p.192.

23. Ibid., p.199.

24. Ibid., p.235.

25. Quoted in *Voluntary Simplicity* by Duane Elgin (New York, William Morrow, 1981), p.72.

26. John Macmurray, *Interpreting the Universe* (London, Faber, 1933), p.109.

27. C.H. Waddington, 'The character of biological form', in L.L. Whyte (ed.) *Aspects of Form* [1951] (London, Lund Humphries, 1968), p.47.

28. Patrick Rivers, *The Stolen Future* (Basingstoke, Green Print, 1988), pp.18–19.

29. Donald P. Hopkins, *Chemicals, Humus, and the Soil* (London, Faber, 1945), p.235.

30. John Macmurray, op. cit., p.110.

31. Aristotle, 'On the Art of Poetry' in T.S. Dorsch (ed.), *Classical Literary Criticism* (Harmondsworth, Penguin, 1965), p.43.

32. Harold Osborne, *Aesthetics and Art Theory* (London, Longmans, 1968), p.193.

33. John Hospers, quoted in G.N. Orsini, 'The ancient roots of a modern idea', in G.S. Rousseau (ed.) *Organic Form* (London, RKP, 1972), p.8.

34. Herbert Read, Preface to first edition of L.L. Whyte (ed.) *Aspects of Form* [1951] (London, Lund Humphries, 1968), p.xxi.

35. Ibid. p.xxii.

36. cf. Ernst Cassirer, *The Philosophy of Symbolic Forms: Vol.1* (New Haven, Yale U.P., 1953), p.154. The Introduction to this volume, by C.W. Hendel, pp.26-32 and p.56, can also be consulted. A study of Goethe's attitude to the beauties of botanical form can be found in Elizabeth Sewell's *The Orphic Voice* (London, RKP, 1960), pp.262–65, and in the essay by Philip C. Ritterbush, 'Organic form—aesthetics and objectivity in the study of form in the life sciences', in *Organic Form*, ed. G.S. Rousseau (London, RKP, 1972).

37. E.W. Sinnott, *The Biology of the Spirit* (London, Gollancz, 1956), p.121.

38. Ernst Haeckel, *Art Forms in Nature* [1904] (New York, Dover Books, 1974).

39. Herbert Read, *The Origins of Form in Art* (London, Thames and Hudson, 1965), p.111.

40. E.I. Watkin, *The Bow in the Clouds* (London, Sheed and Ward, 1931), pp.94-95.

41. John Macmurray, *Religion, Art and Science* (Liverpool U.P., 1961), p.31.

42. Frank Lloyd Wright, *An Organic Architecture* (London, Lund Humphries, 1939), p.3.

43. cf. John Macmurray, *Religion, Art and Science* (Liverpool U.P., 1961) p.44.

44. Ernst Cassirer, *The Philosophy of Symbolic Forms: Vol.3* (New Haven, Yale U.P., 1957), p.38.

45. N.O. Lossky, op. cit., p.63.

46. Ibid. Chapter VI.

47. J.C. Smuts, op. cit. p.353.

48. Wendell Berry, *Home Economics* (San Francisco, North Point Press, 1987), p.ix.

49. Wes Jackson, op. cit. pp.30–31.

50. Vandana Shiva, op. cit., p.40.

51. Philip Sherrard, *The Rape of Man and Nature* (Ipswich, Golgonooza Press, 1987), Chapter 4.

52. cf. Jonathon Porritt and David Winner, *The Coming of the Greens* (London, Fontana, 1988), Chapter 11; and Tim Cooper, *Green Christianity* (London, Spire, 1990).

53. James Lovelock, *The Ages of Gaia* (Oxford U.P., 1988), p.206.

54. Ibid. p.212.

55. Ernst Haeckel, *The Wonders of Life* (London, Watts and Co., 1910), p.111.

56. Rupert Sheldrake, 'Rebirth of Nature', *Resurgence* No. 136 (September/October 1989), p.35.

57. cf. Jeremy Seabrook, *The Myth of the Market* (Hartland, Green Books, 1990).

58. cf. Anna Bramwell, *Ecology in the 20th Century* (New Haven, Yale U.P., 1989), pp. 118–122; and Philip Conford (ed.) *The Organic Tradition* (Hartland, Green Books, 1988), pp.13–15.

59. cf. Philip Conford, op. cit., Chapter 8; and *The Peckham Experiment* by I.H. Pearse and L.H. Crocker (London, Allen and Unwin, 1943).

60. Alison Stallibrass, *Being Me and Also Us* (Edinburgh, Scottish Academic Press, 1989), p.232.

61. Ibid. p.230.

62. Bernard Bosanquet, *The Philosophical Theory of the State* [1899] (London, Macmillan, 1958), pp.227-28.

63. Ernst Haeckel, *The Wonders of Life*, (London, Watts and Co., 1910), pp.134–35.

64. H.E. Goad, *The Making of the Corporate State* (London, Christophers, 1932), p.145.

65. Ibid. p.101.

66. E.I. Watkin, *A Philosophy of Form* [1938] (London, Sheed and Ward, 1950), p.195.

67. cf. Anna Bramwell, op. cit., and Richard Griffiths, *Fellow Travellers of the Right* [1980] (O.U.P., 1983).

68. David Lorimer, on the thought of Sir George Trevelyan, *Resurgence* No. 120 (January/February 1987), p.39.

69. H.J. Massingham, *Remembrance* (London, Batsford, 1942), p.vi.

70. cf. Donald Worster, op. cit. p.285.

71. Wendell Berry, 'Taking Nature's Measure', *Resurgence* No. 142 (September/October 1990), p.20.

72. cf. Peter C. Gould, *Early Green Politics* (Brighton, Harvester Press, 1988).

73. Duane Elgin, op. cit. pp.285-86.

74. N.O. Lossky, op. cit.., p.188.

75. E.I. Watkin, *A Philosophy of Form* [1938] (London, Sheed and Ward, 1950), p.187.

76. For a full discussion of this issue, see Watkin, op. cit., pp.179-205.

77. John Macmurray, *Interpreting the Universe* (London, Faber, 1933), p.121.

INDEX